(£27·50 NEW)

CW00369240

Oxford Lecture Series in
Mathematics and its Applications 1

Series editors

John Ball   Dominic Welsh

Preface

A fundamental problem with quantum theories of gravity, as opposed to the other forces of nature, is that in Einstein's theory of gravity, general relativity, there is no background geometry to work with: the geometry of spacetime itself becomes a dynamical variable. This is loosely summarized by saying that the theory is 'generally covariant'. The standard model of the strong, weak, and electromagnetic forces is dominated by the presence of a finite-dimensional group, the Poincaré group, acting as the symmetries of a spacetime with a fixed geometrical structure. In general relativity, on the other hand, there is an infinite-dimensional group, the group of diffeomorphisms of spacetime, which permutes different mathematical descriptions of any given spacetime geometry satisfying Einstein's equations. This causes two difficult problems. On the one hand, one can no longer apply the usual criterion for fixing the inner product in the Hilbert space of states, namely that it be invariant under the geometrical symmetry group. This is the so-called 'inner product problem'. On the other hand, dynamics is no longer encoded in the action of the geometrical symmetry group on the space of states. This is the so-called 'problem of time'.

The prospects for developing a mathematical framework for quantizing gravity have been considerably changed by a number of developments in the 1980s that at first seemed unrelated. On the one hand, Jones discovered a new polynomial invariant of knots and links, prompting an intensive search for generalizations and a unifying framework. It soon became clear that the new knot polynomials were intimately related to the physics of 2-dimensional systems in many ways. Atiyah, however, raised the challenge of finding a manifestly 3-dimensional definition of these polynomials. This challenge was met by Witten, who showed that they occur naturally in Chern–Simons theory. Chern–Simons theory is a quantum field theory in 3 dimensions that has the distinction of being generally covariant and also a gauge theory, that is a theory involving connections on a bundle over spacetime. Any knot thus determines a quantity called a 'Wilson loop,' the trace of the holonomy of the connection around the knot. The knot polynomials are simply the expectation values of Wilson loops in the vacuum state of Chern–Simons theory, and their diffeomorphism invariance is a consequence of the general covariance of the theory. In the explosion of work that followed, it became clear that a useful framework for understanding this situation is Atiyah's axiomatic description of a 'topological quantum field theory', or TQFT.

On the other hand, at about the same time as Jones' initial discovery, Ashtekar discovered a reformulation of general relativity in terms of what are now called the 'new variables'. This made the theory much more closely resemble a gauge theory. The work of Ashtekar proceeds from the 'canonical' approach to quantization, meaning that solutions to Einstein's equations are identified with their initial data on a given spacelike hypersurface. In this approach the action of the diffeomorphism group gives rise to two constraints on initial data: the diffeomorphism constraint, which generates diffeomorphisms preserving the spacelike surface, and the Hamiltonian constraint, which generates diffeomorphisms that move the surface in a timelike direction. Following the procedure invented by Dirac, one expects the physical states to be annihilated by certain operators corresponding to the constraints.

In an effort to find such states, Rovelli and Smolin turned to a description of quantum general relativity in terms of Wilson loops. In this 'loop representation', they saw that (at least formally) a large space of states annihilated by the constraints can be described in terms of invariants of knots and links. The fact that link invariants should be annihilated by the diffeomorphism constraint is very natural, since a link invariant is nothing other than a function from links to the complex numbers that is invariant under diffeomorphisms of space. In this sense, the relation between knot theory and quantum gravity is a natural one. But the fact that link invariants should also be annihilated by the Hamiltonian constraint is deeper and more intriguing, since it hints at a relationship between knot theory and 4-dimensional mathematics.

More evidence for such a relationship was found by Ashtekar and Kodama, who discovered a strong connection between Chern–Simons theory in 3 dimensions and quantum gravity in 4 dimensions. Namely, in terms of the loop representation, the same link invariant that arises from Chern–Simons theory—a certain normalization of the Jones polynomial called the Kauffman bracket—also represents a state of quantum gravity with cosmological constant, essentially a quantization of anti-deSitter space. The full ramifications of this fact have yet to be explored.

The present volume is the proceedings of a workshop on knots and quantum gravity held on May 14th–16th at the University of California at Riverside, under the auspices of the Departments of Mathematics and Physics. The goal of the workshop was to bring together researchers in quantum gravity and knot theory and pursue a dialog between the two subjects.

On Friday the 14th, Dana Fine began with a talk on 'Chern–Simons theory and the Wess–Zumino–Witten model'. Witten's original work deriving the Jones invariant of links (or, more precisely, the Kauffman bracket) from Chern–Simons theory used conformal field theory in 2 dimensions as a key tool. By now the relationship between conformal field theory and topological quantum field theories in 3 dimensions has been explored from a number of viewpoints. The path integral approach, however, has not yet been worked out in full mathematical rigor. In this talk Fine described work in progress on reducing the Chern–Simons path integral on S^3 to the path integral for the Wess–Zumino–Witten model.

Oleg Viro spoke on 'Simplicial topological quantum field theories'. Recently there has been increasing interest in formulating TQFTs in a manner that relies upon triangulating spacetime. In a sense this is an old idea, going back to the Regge–Ponzano model of Euclidean quantum gravity. However, this idea was given new life by Turaev and Viro, who rigorously constructed the Regge–Ponzano model of 3d quantum gravity as a TQFT based on the $6j$ symbols for the quantum group $SU_q(2)$. Viro discussed a variety of approaches of presenting manifolds as simplicial complexes, cell complexes, etc., and methods for constructing TQFTs in terms of these data.

Saturday began with a talk by Renate Loll on 'The loop formulation of gauge theory and gravity', introduced the loop representation and various mathematical issues associated with it. She also discussed her work on applications of the loop representation to lattice gauge theory. Abhay Ashtekar's talk, 'Loop transforms', largely concerned the paper with Jerzy Lewandowski appearing in this volume. The goal here is to make the loop representation into rigorous mathematics. The notion of measures on the space $\mathcal{A}/\mathcal{G}$ of connections modulo gauge transformations has long been a key concept in gauge theory, which, however, has been notoriously difficult to make precise. A key notion developed by Ashtekar and Isham for this purpose is that of the 'holonomy C*-algebra', an algebra of observables generated by Wilson loops. Formalizing the notion of a measure on $\mathcal{A}/\mathcal{G}$ as a state on this algebra, Ashtekar and Lewandowski have been able to construct such a state with remarkable symmetry properties, in some sense analogous to Haar measure on a compact Lie group. Ashtekar also discussed the implications of this work for the study of quantum gravity.

Pullin spoke on 'The quantum Einstein equations and the Jones polynomial', describing his work with Bernd Brügmann and Rodolfo Gambini on this subject. He also sketched the proof of a new result, that the coefficient of the 2nd term of the Alexander–Conway polynomial represents a state of quantum gravity with zero cosmological constant. His paper with Gambini in this volume is a more detailed proof of this fact.

On Saturday afternoon, Louis Kauffman spoke on 'Vassiliev invariants and the loop states in quantum gravity'. One aspect of the work of Brüegmann, Gambini, and Pullin that is especially of interest to knot theorists is that it involves extending the bracket invariant to generalized links admitting certain kinds of self-intersections. This concept also plays a major role in the study of Vassiliev invariants of knots. Curiously, however, the extensions occurring in the two cases are different. Kauffman explained their relationship from the path integral viewpoint.

Sunday morning began with a talk by Gerald Johnson, 'Introduction to the Feynman integral and Feynman's operational calculus', on his work with Michel Lapidus on rigorous path integral methods. Viktor Ginzburg then spoke on 'Vassiliev invariants of knots', and in the afternoon, Paolo Cotta-Ramusino spoke on '4d quantum gravity and knot theory'. This talk dealt with his work in progress with Maurizio Martellini. Just as Chern–Simons theory gives a great deal of information on knots in 3 dimensions, there appears to be a relationship between a certain class of 4-dimensional theories and so-called '2-knots', that is, embedded surfaces in 4 dimensions. This class, called 'BF theories', includes quantum gravity in the Ashtekar formulation, as well as Donaldson theory. Cotta-Ramusino and Martellini have described a way to construct observables associated to 2-knots, and are endeavoring to prove at a perturbative level that they give 2-knot invariants.

Louis Crane spoke on 'Quantum gravity, spin geometry, and categorical physics'. This was a review of his work on the relation between Chern–Simons theory and 4-dimensional quantum gravity, his construction with David Yetter of a 4-dimensional TQFT based on the $15j$ symbols for $SU_q(2)$, and his work with Igor Frenkel on certain braided tensor 2-categories. In the final talk of the workshop, John Baez spoke on 'Strings, loops, knots and gauge fields'. He attempted to clarify the similarity between the loop representation of quantum gravity and string theory. At a fixed time both involve loops or knots in space, but the string-theoretic approach is also related to the study of 2-knots.

As some of the speakers did not submit papers for the proceedings, papers were also solicited from Steve Carlip and also from J. Scott Carter and Masahico Saito. Carlip's paper, 'Geometric structures and loop variables', treats the thorny issue of relating the loop representation of quantum gravity to the traditional formulation of gravity in terms of a metric. He treats quantum gravity in 3 dimensions, which is an exactly soluble test case. The paper by Carter and Saito, 'Knotted surfaces, braid movies, and beyond', contains a review of their work on 2-knots as well as a number of new results on 2-braids. They also discuss the role in 4-dimensional topology of new algebraic structures such as braided tensor 2-categories.

The editor would like to thank many people for making the workshop a success. First and foremost, the speakers and other participants are to be congratulated for making it such a lively and interesting event. The workshop was funded by the Departments of Mathematics and Physics of U. C. Riverside, and the chairs of these departments, Albert Stralka and Benjamin Shen, were crucial in bringing this about. Michel Lapidus deserves warm thanks for his help in planning the workshop. Invaluable help in organizing the workshop and setting things up was provided by the staff of the Department of Mathematics, and particularly Susan Spranger, Linda Terry, and Chris Truett. Arthur Greenspoon kindly volunteered to help edit the proceedings. Lastly, thanks go to all the participants in the

Knots and Quantum Gravity Seminar at U. C. Riverside, and especially Jim Gilliam, Javier Muniain, and Mou Roy, who helped set things up.

Contents

Contents

The Loop Formulation of Gauge Theory and Gravity

Renate Loll

Center for Gravitational Physics and Geometry,
Pennsylvania State University,
University Park, Pennsylvania 16802, USA
(email: loll@phys.psu.edu)

Abstract

This chapter contains an overview of the loop formulation of Yang–Mills theory and 3+1-dimensional gravity in the Ashtekar form. Since the configuration spaces of these theories are spaces of gauge potentials, their classical and quantum descriptions may be given in terms of gauge-invariant Wilson loops. I discuss some mathematical problems that arise in the loop formulation and illustrate parallels and differences between the gauge theoretic and gravitational applications. Some remarks are made on the discretized lattice approach.

1 Introduction

This chapter summarizes the main ideas and basic mathematical structures that are necessary to understand the loop formulation of both gravity and gauge theory. It is also meant as a guide to the literature, where all the relevant details may be found. Many of the mathematical problems arising in the loop approach are described in a related review article [21]. I have tried to treat Yang–Mills theory and gravity in parallel, but also to highlight important differences. Section 2 summarizes the classical and quantum features of gauge theory and gravity, formulated as theories of connections. In Section 3, the Wilson loops and their properties are introduced, and Section 4 discusses the classical equivalence between the loop and the connection formulation. Section 5 contains a summary of progress in a corresponding lattice loop approach, and in Section 6 some of the main ingredients of quantum loop representations are discussed.

2 Yang–Mills theory and general relativity as dynamics on connection space

In this section I will briefly recall the Lagrangian and Hamiltonian formulation of both pure Yang–Mills gauge theory and gravity, defined on a 4-dimensional manifold $M = \Sigma^3 \times \mathbf{R}$ of Lorentzian signature. The gravitational theory will be described in the Ashtekar form, in which it most closely resembles a gauge theory.

The classical actions defining these theories are (up to overall constants)

$$S_{YM}[^4A] = -\int_M d^4x \, \mathrm{Tr}\, F_{\mu\nu}F^{\mu\nu} = -\int_M \mathrm{Tr}(F \wedge *F) \qquad (2.1)$$

and

$$S_{GR}[^4A, e] = \int_M d^4x \, e \, e_I^\mu e_J^\nu F_{\mu\nu}{}^{IJ} = \int_M \epsilon_{IJKL}\, e^I \wedge e^J \wedge F^{KL} \qquad (2.2)$$

for Yang–Mills theory and general relativity respectively. In (2.1), the basic variable A is a Lie(G)-valued connection 1-form on M, where G denotes the compact and semisimple Lie-structure group of the Yang–Mills theory (typically $G = SU(N)$). In (2.2), which is a first-order form of the gravitational action, A is a self-dual, $so(3,1)_{\mathbf{C}}$-valued connection 1-form, and e a vierbein, defining an isomorphism of vector spaces between the tangent space of M and the fixed internal space with the Minkowski metric η_{IJ}. The connection A is self-dual in the internal space, $A_\mu^{MN} = -\frac{i}{2}\epsilon^{MN}{}_{IJ}A_\mu^{IJ}$, with the totally antisymmetric ϵ-tensor. (A mathematical characterization of the action (2.2) is contained in Baez' chapter [9].) As usual, F denotes in both cases the curvature associated with the 4-dimensional connection ^4A.

Note that the action (2.2) has the unusual feature of being complex. I will not explain here why this still leads to a well-defined theory equivalent to the standard formulation of general relativity. Details may be found in [3, 4].

Our main interest will be the Hamiltonian formulations associated with (2.1) and (2.2). Assuming the underlying principal fibre bundles $P(\Sigma, G)$ to be trivial, we can identify the relevant phase spaces with cotangent bundles over affine spaces $\mathcal{A}$ of Lie(G)-valued connections on the 3-dimensional manifold Σ (assumed compact and orientable). The cotangent bundle $T^*\mathcal{A}$ is coordinatized by canonically conjugate pairs $(A, \tilde{E})$, where A is a Lie(G)-valued, pseudo-tensorial 1-form ('gauge potential') and $\tilde{E}$ a Lie(G)-valued vector density ('generalized electric field') on Σ. We have $G = SU(N)$, say, for gauge theory, and $G = SO(3)_{\mathbf{C}}$ for gravity. The standard symplectic structure on $T^*\mathcal{A}$ is expressed by the fundamental Poisson bracket relations

$$\{A_a^i(x), \tilde{E}_j^b(y)\} = \delta_j^i \delta_a^b \delta^3(x - y) \qquad (2.3)$$

for gauge theory, and

$$\{A_a^i(x), \tilde{E}_j^b(y)\} = i\, \delta_j^i \delta_a^b \delta^3(x - y) \qquad (2.4)$$

for gravity, where the factor of i arises because the connection A is complex valued. In (2.3) and (2.4), a and b are 3-dimensional spatial indices and i

and j internal gauge algebra indices.

However, points in $T^*\mathcal{A}$ do not yet correspond to physical configurations, because both theories possess gauge symmetries, which are related to invariances of the original Lagrangians under certain transformations on the fundamental variables. At the Hamiltonian level, this manifests itself in the existence of so-called first-class constraints. These are sets $\mathcal{C}$ of functions C on phase space, which are required to vanish, $C = 0$, for physical configurations and at the same time generate transformations between physically indistinguishable configurations. In our case, the first-class constraints for the two theories are

$$\mathcal{C}_{YM} = \{D_a \tilde{E}^{ai}\} \tag{2.5}$$

and

$$\mathcal{C}_{GR} = \{D_a \tilde{E}^{ai}, \; C_a = F_{ab}{}^i \tilde{E}^b_i, \; C = \epsilon^{ijk} F_{ab\,i} \tilde{E}^a_j \tilde{E}^b_k\}. \tag{2.6}$$

The physical interpretation of these constraints is as follows: the so-called Gauss law constraints $D_a \tilde{E}^{ai} = 0$, which are common to both theories, restrict the allowed configurations to those whose covariant divergence vanishes, and at the same time generate local gauge transformations in the internal space. The well-known Yang–Mills transformation law corresponding to a group element $g \in \mathcal{G}$, the set of G-valued functions on Σ, is

$$(A, \tilde{E}) \mapsto (g^{-1}Ag + g^{-1}dg, g^{-1}\tilde{E}g). \tag{2.7}$$

The second set, $\{C_a\}$, of constraints in (2.6) generates 3-dimensional diffeomorphisms of Σ and the last expression, the so-called scalar or Hamiltonian constraint C, generates phase space transformations that are interpreted as corresponding to the time evolution of the spatial slice Σ in M in a spacetime picture. Comparing (2.5) and (2.6), we see that in the Hamiltonian formulation, pure gravity may be interpreted as a Yang–Mills theory with gauge group $G = SO(3)_{\mathbf{C}}$, subject to four additional constraints in each point of Σ. However, the dynamics of the two theories is different; we have

$$H_{YM}(A, \tilde{E}) = \int_\Sigma d^3x \; g_{ab}\left(\frac{1}{\sqrt{g}}\tilde{E}^{ai}\tilde{E}^b_i + \sqrt{g}B^{ai}B^b_i\right) \tag{2.8}$$

as the Hamiltonian for Yang–Mills theory (where for convenience we have introduced the generalized magnetic field $B^a_i := -\frac{1}{2}\epsilon^{abc}F_{bc\,i}$). Note the explicit appearance of a Riemannian background metric g on Σ, to contract indices and ensure the integrand in (2.8) is a density. On the other hand, no such additional background structure is necessary to make the gravitational Hamiltonian well defined. The Hamiltonian H_{GR} for gravity is a sum of a subset of the constraints (2.6),

$$H_{GR}(A, \tilde{E}) = i \int_{\Sigma} d^3x \left(N^a F_{ab}{}^i \tilde{E}_i^b - \frac{i}{2} \underset{\sim}{N} \epsilon^{ijk} F_{ab\,k} \tilde{E}_i^a \tilde{E}_j^b \right), \qquad (2.9)$$

and therefore vanishes on physical configurations. ($\underset{\sim}{N}$ and N^a in (2.9) are Lagrange multipliers, the so-called lapse and shift functions.) This latter feature is peculiar to generally covariant theories, whose gauge group contains the diffeomorphism group of the underlying manifold. A further difference from the gauge-theoretic application is the need for a set of reality conditions for the gravitational theory, because the connection A is a priori a *complex* coordinate on phase space.

The space of physical configurations of Yang–Mills theory is the quotient space $\mathcal{A}/\mathcal{G}$ of gauge potentials modulo local gauge transformations. It is non-linear and has non-trivial topology. After excluding the reducible connections from $\mathcal{A}$, this quotient can be given the structure of an infinite-dimensional manifold for compact and semisimple G [2]. The corresponding physical phase space is then its cotangent bundle $T^*(\mathcal{A}/\mathcal{G})$. Elements of $T^*(\mathcal{A}/\mathcal{G})$ are called *classical observables* of the Yang–Mills theory.

The non-linearities of both the equations of motion and the physical configuration/phase spaces, and the presence of gauge symmetries for these theories, lead to numerous difficulties in their classical and quantum description. For example, in the path integral quantization of Yang–Mills theory, a gauge-fixing term has to be introduced in the 'sum over all configurations'

$$Z = \int_{\mathcal{A}} [dA] \, e^{iS_{YM}}, \qquad (2.10)$$

to ensure that the integration is only taken over one member [4]A of each gauge equivalence class. However, because the principal bundle $\mathcal{A} \to \mathcal{A}/\mathcal{G}$ is non-trivial, we know that a unique and attainable gauge choice does not exist (this is the assertion of the so-called Gribov ambiguity). In any case, since the expression (2.2) can be made meaningful only in a weak-field approximation where one splits $S_{YM} = S_{\text{free}} + S_I$, with S_{free} being just quadratic in A, this approach has not yielded enough information about other sectors of the theory, where the fields cannot be assumed weak.

Similarly, in the canonical quantization, since no good explicit description of $T^*(\mathcal{A}/\mathcal{G})$ is available, one quantizes the theory 'à la Dirac'. This means that one first 'quantizes' on the unphysical phase space $T^*\mathcal{A}$, 'as if there were no constraints'. That is, one uses a formal operator representation of the canonical variable pairs $(A, \tilde{E})$, satisfying the canonical commutation relations

$$[\hat{A}_a^i(x), \hat{\tilde{E}}_j^b(y)] = i\hbar \delta_j^i \delta_a^b \delta^3(x - y), \qquad (2.11)$$

the quantum analogues of the Poisson brackets (2.3). Then a subset of physical wave functions $\mathcal{F}_{\mathrm{phy}}^{YM} = \{\Psi_{\mathrm{phy}}(A)\}$ is projected out from the space of all wave functions, $\mathcal{F} = \{\Psi(A)\}$, according to

$$\mathcal{F}_{\mathrm{phy}}^{YM} = \{\Psi(A) \in \mathcal{F} \,|\, (\widehat{DE})\Psi(A) = 0\}, \tag{2.12}$$

i.e. $\mathcal{F}_{\mathrm{phy}}^{YM}$ consists of all functions that lie in the kernel of the quantized Gauss constraint, $\widehat{DE}$. (For a critical appraisal of the use of such 'Dirac conditions', see [20].) The notations $\mathcal{F}$ and $\mathcal{F}_{\mathrm{phy}}^{YM}$ indicate that these are just spaces of complex-valued functions on $\mathcal{A}$ and do not yet carry any Hilbert space structure. Whereas this is not required on the unphysical space $\mathcal{F}$, one does need such a structure on $\mathcal{F}_{\mathrm{phy}}^{YM}$ in order to make physical statements, for instance about the spectrum of the quantum Hamiltonian $\hat{H}$. Unfortunately, we lack an appropriate scalar product, i.e. an appropriate gauge-invariant measure $[dA]_{\mathcal{G}}$ in

$$\langle \Psi_{\mathrm{phy}}, \Psi'_{\mathrm{phy}} \rangle = \int_{\mathcal{A}/\mathcal{G}} [dA]_{\mathcal{G}} \, \Psi^*_{\mathrm{phy}}(A)\Psi'_{\mathrm{phy}}(A). \tag{2.13}$$

(More precisely, we expect the integral to range over an extension of the space $\mathcal{A}/\mathcal{G}$ which includes also distributional elements [5].) Given such a measure, the space of physical wave functions would be the Hilbert space $\mathcal{H}^{YM} \subset \mathcal{F}_{\mathrm{phy}}^{YM}$ of functions square-integrable with respect to the inner product (2.13).

Unfortunately, the situation for gravity is not much better. Here the space of physical wave functions is the subspace of $\mathcal{F}$ projected out according to

$$\mathcal{F}_{\mathrm{phy}}^{GR} = \{\Psi(A) \in \mathcal{F} \,|\, (\widehat{DE})\Psi(A) = 0, \ \hat{C}_a\Psi(A) = 0, \ \hat{C}\Psi(A) = 0\}. \tag{2.14}$$

Again, one now needs an inner product on the space of such states. Assuming that the states Ψ_{phy} of quantum gravity can be described as the set of complex-valued functions on some (still infinite-dimensional) space $\mathcal{M}$, the analogue of (2.13) reads

$$\langle \Psi_{\mathrm{phy}}, \Psi'_{\mathrm{phy}} \rangle = \int_{\mathcal{M}} [dA] \, \Psi^*_{\mathrm{phy}}(A)\Psi'_{\mathrm{phy}}(A), \tag{2.15}$$

where the measure $[dA]$ is now both gauge and diffeomorphism invariant and projects in an appropriate way to $\mathcal{M}$. If we had such a measure (again, we do not), the Hilbert space $\mathcal{H}^{GR}$ of states for quantum gravity would consist of all those elements of $\mathcal{F}_{\mathrm{phy}}^{GR}$ which are square-integrable with respect to that measure. In the—up to now—absence of explicit measures to make (2.13, 2.15) well defined, our confidence in the procedure outlined here stems from the fact that it can be made meaningful for lower-dimensional model systems (such as Yang–Mills theory in 1+1 and gravity in 2+1 dimensions), where the analogues of the space $\mathcal{M}$ are finite dimensional (cf. also [9]).

Note that there is a very important difference in the role played by the quantum Hamiltonians in gauge theory and gravity. In Yang–Mills theory, one searches for solutions in $\mathcal{H}^{YM}$ of the eigenvector equation $\hat{H}_{YM}\Psi_{\text{phy}}(A) = E\Psi_{\text{phy}}(A)$, whereas in gravity the quantum Hamiltonian $\hat{C}$ is one of the constraints and therefore a physical state of quantum gravity must lie in its kernel, $\hat{C}\Psi_{\text{phy}}(A) = 0$.

After all that has been said about the difficulties of finding a canonical quantization for gauge theory and general relativity, the reader may wonder whether one can make any statements at all about their quantum theories. For the case of Yang–Mills theory, the answer is of course in the affirmative. Many qualitative and quantitative statements about quantum gauge theory can be derived in a regularized version of the theory, where the flat background space(time) is approximated by a hypercubic lattice. Using refined computational methods, one then tries to extract (for either weak or strong coupling) properties of the continuum theory, for example an approximate spectrum of the Hamiltonian $\hat{H}_{YM}$, in the limit as the lattice spacing tends to zero. Still there are non-perturbative features, the most important being the confinement property of Yang–Mills theory, which also in this framework are not well understood.

Similar attempts to discretize the (Hamiltonian) theory have so far been not very fruitful in the treatment of gravity, because, unlike in Yang–Mills theory, there is no fixed background metric (with respect to which one could build a hypercubic lattice), without violating the diffeomorphism invariance of the theory. However, there are more subtle ways in which discrete structures arise in canonical quantum gravity. An example of this is given by the lattice-like structures used in the construction of diffeomorphism-invariant measures on the space $\mathcal{A}/\mathcal{G}$ [6, 8]. One important ingredient we will be focusing on are the non-local loop variables on the space $\mathcal{A}$ of connections, which will be the subject of the next section.

3   Introducing loops!

A *loop* in Σ is a continuous map γ from the unit interval into Σ,

$$
\begin{aligned}
\gamma : [0,1] &\;\rightarrow\; \Sigma \\
s &\;\mapsto\; \gamma^a(s), \\
\text{s. t. } \gamma(0) &\;=\; \gamma(1).
\end{aligned}
\tag{3.1}
$$

The set of all such maps will be denoted by $\Omega\Sigma$, the loop space of Σ. Given such a loop element γ, and a space $\mathcal{A}$ of connections, we can define a complex function on $\mathcal{A} \times \Omega\Sigma$, the so-called *Wilson loop*,

$$T_A(\gamma) := \frac{1}{N} \mathrm{Tr}_R \, \mathrm{P} \exp \oint_\gamma A. \qquad (3.2)$$

The path-ordered exponential of the connection A along the loop γ, $U_\gamma = \mathrm{P} \exp \oint_\gamma A$, is also known as the holonomy of A along γ. The holonomy measures the change undergone by an internal vector when parallel transported along γ. The trace is taken in the representation R of G, and N is the dimensionality of this linear representation. The quantity $T_A(\gamma)$ was first introduced by K. Wilson as an indicator of confining behaviour in the lattice gauge theory [32]. It measures the curvature (or field strength) in a gauge-invariant way (the components of the tensor F_{ab} itself are not measurable), as can easily be seen by expanding $T_A(\gamma)$ for an infinitesimal loop γ. The Wilson loop has a number of interesting properties.

1. It is invariant under the local gauge transformations (2.7), and therefore an observable for Yang–Mills theory.
2. It is invariant under orientation-preserving reparametrizations of the loop γ.
3. It is independent of the basepoint chosen on γ.

It follows from 1 and 2 that the Wilson loop is a function on a cross-product of the quotient spaces, $T_A(\gamma) \in \mathcal{F}(\mathcal{A}/\mathcal{G} \times \Omega\Sigma/\mathrm{Diff}^+(S^1))$.

In gravity, the Wilson loops are not observables, since they are not invariant under the full set of gauge transformations of the theory. However, there is now one more reason to consider variables that depend on closed curves in Σ, since observables in gravity also have to be invariant under diffeomorphisms of Σ. This means going one step further in the construction of such observables, by considering Wilson loops that are constant along the orbits of the $\mathrm{Diff}(\Sigma)$-action,

$$T(\varphi \circ \gamma) = T(\gamma), \ \forall \varphi \in \mathrm{Diff}(\Sigma). \qquad (3.3)$$

Such a loop variable will just depend on what is called the generalized knot class $K(\gamma)$ of γ ('generalized' since γ may possess self-intersections and self-overlaps, which is not usually considered in knot theory). Note that a similar construction for taking care of the diffeomorphism invariance is not very contentful if we start from a *local* scalar function, $\phi(x)$, say. The analogue of (3.3) would then tell us that ϕ had to be constant on any connected component of Σ. The set of such variables is not big enough to account for all the degrees of freedom of gravity. In contrast, loops can carry diffeomorphism-invariant information such as their number of self-intersections, number of points of non-differentiability, etc.

The use of non-local holonomy variables in gauge field theory was pioneered by Mandelstam in his 1962 paper on 'Quantum electrodynamics without potentials' [24], and later generalized to the non-Abelian theory by Bialynicki-Birula [10] and again Mandelstam [25]. Another upsurge in interest in path-dependent formulations of Yang–Mills theory took place at the end of the 1970s, following the observation of the close formal resemblance of a particular form of the Yang–Mills equations in terms of holonomy variables with the equations of motion of the 2-dimensional nonlinear sigma model [1, 28]. Another set of equations for the Wilson loop was derived by Makeenko and Migdal (see, for example, the review [26]).

The hope underlying such approaches has always been that one may be able to find a set of basic variables for Yang–Mills theory which are better suited to its quantization than the local gauge potentials $A(x)$. Owing to their simple behaviour under local gauge transformations, the holonomy or the Wilson loop seem like ideal candidates. The aim has therefore been to derive suitable differential equations in loop space for the holonomy U_γ, its trace, $T(\gamma)$, or for their vacuum expectation values, which are to be thought of as the analogues of the usual local Yang–Mills equations of motion. The fact that none of these attempts have been very successful can be attributed to a number of reasons.

- To make sense of differential equations on loop space, one first has to introduce a suitable topology, and then set up a differential calculus on $\Omega\Sigma$. Even if we give $\Omega\Sigma$ locally the structure of a topological vector space, and are able to define differentiation, this in general will ensure the existence of neither inverse and implicit function theorems nor theorems on the existence and uniqueness of solutions of differential equations. Such issues have only received scant attention in the past.

- Since the space of all loops in Σ is so much larger than the set of all points in Σ, one expects that not all loop variables will be independent. This expectation is indeed correct, as will be explained in the next section. Still, the ensuing redundancy is hard to control in the continuum theory, and has therefore obscured many attempts of establishing an equivalence between the connection and the loop formulation of gauge theory. For example, there is no action principle in terms of loop variables, which leaves considerable freedom for deriving 'loop equations of motion'. As another consequence, perturbation theory in the loop approach always had to fall back on the perturbative expansion in terms of the connection variable A.

4 Equivalence between the connection and loop formulations

In this section I will discuss the motivation for adopting a 'pure loop formulation' for any theory whose configuration space comes from a space $\mathcal{A}/\mathcal{G}$ of connection 1-forms modulo local gauge transformations. For the Yang–Mills theory itself, there is a strong physical motivation to choose the variables $T(\gamma)$ as a set of basic variables: on physical grounds, one requires only physical observables O (i.e. in our case the Wilson loops, and not the local fields A) to be represented by self-adjoint operators $\hat{O}$ in the quantum theory, with the classical Poisson relations $\{O, O'\}$ going over to the quantum commutators $i\hbar[\hat{O}, \hat{O}']$.

There is also a mathematical justification for re-expressing the classical theory in terms of loop variables. It is a well-known result that one can reconstruct the gauge potential A up to gauge transformations from the knowledge of all holonomies U_γ, for any structure group $G \subset GL(N, \mathbf{C})$. Therefore all that remains to be shown is that one can reconstruct the holonomies $\{U_\gamma\}$ from the set of Wilson loops $\{T(\gamma)\}$. It turns out that, at least for compact G, there is indeed an equivalence between the two quotient spaces [17]

$$\frac{\mathcal{A}}{\mathcal{G}} \simeq \frac{\{T(\gamma),\, \gamma \in \Omega\Sigma/\mathrm{Diff}^+(S^1)\}}{\text{Mandelstam constraints}}. \tag{4.1}$$

By Mandelstam constraints I mean a set of necessary and sufficient conditions on generic complex functions $f(\gamma)$ on $\Omega\Sigma/\mathrm{Diff}^+(S^1)$ that ensure they are the traces of holonomies of a group G in a given representation R. They are typically algebraic, non-linear constraints among the $T(\gamma)$. To obtain a complete set of such conditions (without making reference to the connection space $\mathcal{A}$) is a non-trivial problem that for general G to my knowledge has not been solved.

For example, for $G = SL(2, \mathbf{C})$ in the fundamental representation by 2×2-matrices, the Mandelstam constraints are

(a) $$T(\text{point loop}) = 1$$
(b) $$T(\gamma) = T(\gamma^{-1})$$
(c) $$T(\gamma_1 \circ \gamma_2) = T(\gamma_2 \circ \gamma_1) \tag{4.2}$$
(d) $$T(\gamma) = T(\gamma'), \quad \gamma' = \gamma \circ \mu \circ \mu^{-1}$$
(e) $$T(\gamma_1)T(\gamma_2) = \tfrac{1}{2}\Big(T(\gamma_1 \circ \lambda \circ \gamma_2 \circ \lambda^{-1}) + T(\gamma_1 \circ \lambda \circ \gamma_2^{-1} \circ \lambda^{-1})\Big).$$

The loop γ' in eqn (d) is the loop γ with an 'appendix' $\mu \circ \mu^{-1}$ attached, where μ is any path (not necessarily closed) starting at a point of γ (see Fig. 1). In (e), λ is a path starting at a point of γ_1 and ending at γ_2 (see Fig. 2). The circle $\circ$ denotes the usual path composition. It is believed that the set (4.2) and all conditions that can be derived from (4.2) exhaust the Mandelstam constraints for this particular gauge group and representation, although I am not aware of a formal proof. Still, owing to the non-compactness of G, it can be shown that not every element of $\mathcal{A}/\mathcal{G}$ can be reconstructed from the Wilson loops. This is the case for holonomies that take their values in the so-called subgroup of null rotations [18]. However, this leads only to an incompleteness 'of measure zero' of the Wilson

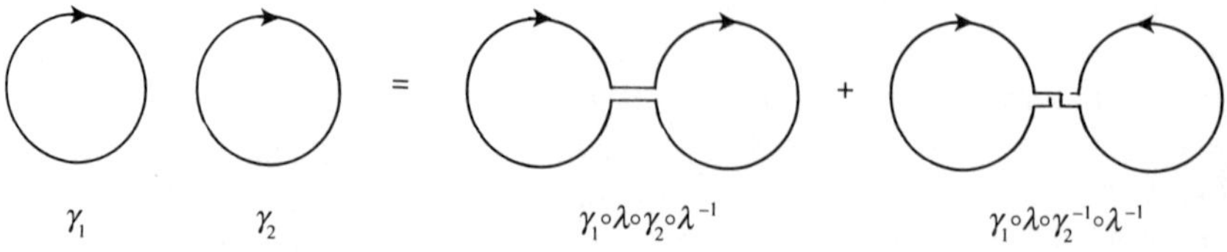

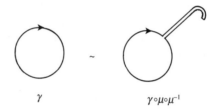

$$\gamma_1 \qquad\qquad \gamma_2 \qquad\qquad\qquad \gamma_1 \circ \lambda \circ \gamma_2 \circ \lambda^{-1} \qquad\qquad \gamma_1 \circ \lambda \circ \gamma_2^{-1} \circ \lambda^{-1}$$

FIG. 1. Adding an appendix $\mu \circ \mu^{-1}$ to the loop γ does not change the Wilson loop

$$\gamma \qquad\qquad\qquad \gamma \circ \mu \circ \mu^{-1}$$

FIG. 2. The loop configurations related by the Mandelstam constraint
$$T(\gamma_1)T(\gamma_2) = \tfrac{1}{2}\Big(T(\gamma_1 \circ \lambda \circ \gamma_2 \circ \lambda^{-1}) + T(\gamma_1 \circ \lambda \circ \gamma_2^{-1} \circ \lambda^{-1})\Big)$$

loops on $\mathcal{A}/\mathcal{G}$.

If one wants to restrict the group G to a subgroup of $SL(2,\mathbf{C})$, there are further conditions on the loop variables, which now take the form of inequalities [23]. For $SU(2) \subset SL(2,\mathbf{C})$, one finds the two conditions

$$-1 \leq T(\gamma) \leq 1,$$
$$\Big(T(\gamma_1 \circ \gamma_2) - T(\gamma_1 \circ \gamma_2^{-1})\Big)^2 \leq 4(1 - T(\gamma_1)^2)(1 - T(\gamma_2)^2), \quad (4.3)$$

whereas for $SU(1,1) \subset SL(2,\mathbf{C})$ one derives

$$T(\gamma_1)^2 \leq 1 \ \wedge \ T(\gamma_2)^2 \leq 1 \Rightarrow$$
$$\Big(T(\gamma_1 \circ \gamma_2) - T(\gamma_1 \circ \gamma_2^{-1})\Big)^2 \geq 4(1 - T(\gamma_1)^2)(1 - T(\gamma_2)^2), \quad (4.4)$$

and no restrictions for other values of $T(\gamma_1)$ and $T(\gamma_2)$. Unfortunately,

(4.3) and (4.4) do not exhaust all inequalities there are for $SU(2)$ and $SU(1,1)$ respectively. There are more inequalities corresponding to more complicated configurations of more than two intersecting loops, which are critically dependent on the geometry of the loop configuration.

Let me finish this section with some further remarks on the Mandelstam constraints:

- In cases where the Wilson loops form a complete set of variables, they are actually overcomplete, due to the existence of the Mandelstam constraints (i.e. implicitly the non-trivial topological structure of $\mathcal{A}/\mathcal{G}$). This is the overcompleteness I referred to in the previous section.

- Note that one may interpret some of the constraints (4.2) as constraints on the underlying loop space, i.e. the Wilson loops are effectively not functions on $\Omega\Sigma/\text{Diff}^+(S^1)$, but on some appropriate quotient with respect to equivalence relations such as $\gamma \sim \gamma \circ \mu \circ \mu^{-1}$. The existence of such relations makes it difficult to apply directly mathematical results on the loop space $\Omega\Sigma$ or the space $\Omega\Sigma/\text{Diff}^+(S^1)$ of oriented, unparametrized loops (such as [12, 13, 14, 31]) to the present case.

- One has to decide how the Mandelstam constraints are to be carried over to the quantum theory. One may try to eliminate as many of them as possible already in the classical theory, but real progress in this direction has only been made in the discretized lattice gauge theory (see Section 5 below). Another possibility will be mentioned in Section 6.

- Finally, let me emphasize that the equivalence of the connection and the loop description at the classical level does not necessarily mean that quantization with the Wilson loops as a set of basic variables will be equivalent to a quantization in which the connections A are turned into fundamental quantum operators. One may indeed hope that at some level, those two possibilities lead to different results, with the loop representation being better suited for the description of the non-perturbative structure of the theory.

5   Some lattice results on loops

This section describes the idea of taking the Wilson loops as the set of basic variables, as applied to gauge theory, although some of the techniques and conclusions may be relevant for gravitational theories too. The lattice discretization is particularly suited to the loop formulation, because in it the gauge fields are taken to be concentrated on the 1-dimensional links l_i of the lattice. The Wilson loops are easily formed by multiplying together the holonomies $U_{l_1} \ldots U_{l_n}$ of a set of contiguous, oriented links that form

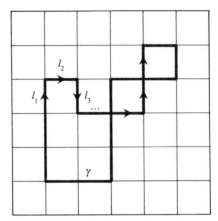

FIG. 3. γ is a loop on the hypercubic lattice N^d, $N = 6$, $d = 2$.

a closed chain on the lattice (see Fig. 3),

$$T(\gamma) = \text{Tr}_R\, U_{l_1} U_{l_2} \ldots U_{l_n}. \tag{5.1}$$

We take the lattice to be the hypercubic lattice N^d, with periodic boundary conditions (N is the number of links in each direction), and the gauge group G to be $SU(2)$, again in the fundamental representation. Although the lattice is finite, there are an infinite number of closed contours one can form on it, and hence an infinite number of Wilson loops. At the same time, the number of Mandelstam constraints (4.2) is infinite. The only constraints that cause problems are those of type (e), because of their non-linear and highly coupled character. Nevertheless, as I showed in [19], it is possible to solve this coupled set of equations and obtain an explicit local description of the finite-dimensional physical configuration space

$$\mathcal{C}_{\text{phy}} = \left.\frac{\mathcal{A}}{\mathcal{G}}\right|_{\text{lattice}} = \frac{\{T(\gamma),\ \gamma \text{ a lattice loop}\}}{\text{Mandelstam constraints}} \tag{5.2}$$

in terms of an independent set of Wilson loops. The final solution for the independent degrees of freedom is surprisingly simple: it suffices to use the Wilson loops of small lattice loops, consisting of no more than six lattice links. At this point it may be noted that a similar problem arises in 2+1 gravity (in the connection formulation) on a manifold $\Sigma^g \times \mathbf{R}$, with a compact, 2-dimensional Riemann surface Σ^g of genus g. One also has a discrete set of loops, namely, the elements of the homotopy group of Σ^g, and can introduce the corresponding Wilson loops. Again, one is interested in finding a maximally reduced set of such Wilson loops, in terms of which all the others can be expressed, using the Mandelstam constraints (see, for example, [7, 27]).

Of course it does not suffice to establish a set of independent loop variables on the lattice; one also has to show that the dynamics can be expressed in a manageable form in terms of these variables. Some important problems are the following.

1. Since the space $\mathcal{C}_{\mathrm{phy}} = \mathcal{A}/\mathcal{G}$ is topologically non-trivial, the set of independent lattice Wilson loops cannot provide a good global chart for $\mathcal{C}_{\mathrm{phy}}$. It would therefore be desirable to find a minimal embedding into a space of loop variables that *is* topologically trivial (together with a finite number of non-linear constraints that define the space $\mathcal{C}_{\mathrm{phy}}$).

2. What is the explicit form of the Jacobian of the transformation from the holonomy link variables to the Wilson loops? From this one could derive an effective action/Hamiltonian for gauge theory, and maybe derive a prescription of how to translate local quantities into their loop space analogues.

3. Is it possible to define a perturbation theory in loop space, once one understands which Wilson loop configurations contribute most in the path integral, say? This would be an interesting alternative to the usual perturbation theory in A or the $\frac{1}{N}$-expansion, and may lead to new and more efficient ways of performing calculations in lattice gauge theory.

4. Lastly, in finding a solution to the Mandelstam constraints it was crucial to have the concept of a smallest loop size on the lattice (the loops of link length 4 going around a single plaquette). What are the consequences of this result for the continuum theory, where there is no obstruction to shrinking loops down to points?

Some progress on points 1 and 2 has been made for small lattices [22], and it is clear that all of the above-posed problems are rather non-trivial, even in the discretized lattice theory where the number of degrees of freedom is effectively finite. One may be able to apply some of the techniques used here to 3+1-dimensional gravity, where discrete structures appear too (the generalized knot classes) after factoring out by the spatial diffeomorphisms.

6   Quantization in the loop approach

I will now return to the discussion of the continuum theory of both gravity and gauge theory, and review a few concepts that have been used in their quantization in a loop formulation. The contributions by Ashtekar and Lewandowski [6] and Pullin [29] in this volume contain more details about some of the points raised here.

The main idea is to base the canonical quantization of a theory with configuration space $\mathcal{A}$ on the non-local, gauge-invariant phase space variables $T(\gamma)$ (and not on some gauge-covariant local field variables). This is a somewhat unusual procedure in the quantization of a field theory, but takes directly into account the non-linearities of the theory.

You may have noted that so far we have only talked about *configuration* space variables, the Wilson loops. For a canonical quantization we obviously need some momentum variables that depend on the generalized electric field $\tilde{E}$. One choice of a generalized Wilson loop that also depends on $\tilde{E}$ is

$$T^a_{(A,\tilde{E})}(\gamma, s) := \text{Tr } U_\gamma(s)\tilde{E}^a(\gamma(s)). \tag{6.1}$$

This variable depends now on both a loop and a marked point, and it is still gauge invariant under the transformations (2.7). Also it is strictly speaking not a function on phase space, but a vector density. This latter fact may be remedied by integrating T^a over a 2-dimensional 'ribbon' or 'strip', i.e. a non-degenerate 1-parameter congruence of curves $\gamma_t(s) =: R(s,t)$, $t \in [0,1]$ [5]. In any case, the 'momentum Wilson loop' depends not just on a loop, but needs an additional geometric ingredient. I will use here the unsmeared version since it does not affect what I am going to say.

The crucial ingredient in the canonical quantization is the fact that the loop variables $(T(\gamma), T^a(\gamma, s))$ form a closing algebra with respect to the canonical symplectic structure on $T^*\mathcal{A}$. This fact was first established by Gambini and Trias [16] for the gauge group $SU(N)$ and later rediscovered by Rovelli and Smolin in the context of gravity [30]. For the special case of $G = SL(2, \mathbf{C})$ (or any of its subgroups) in the fundamental representation, the Poisson algebra of the loop variable s is

$$\{T(\gamma_1), T(\gamma_2)\} = 0,$$
$$\{T^a(\gamma_1, s), T(\gamma_2)\} = -\Delta^a(\gamma_2, \gamma_1(s))\left(T(\gamma_1 \circ_s \gamma_2) - T(\gamma_1 \circ_s \gamma_2^{-1})\right),$$
$$\{T^a(\gamma_1, s), T^b(\gamma_2, t)\} = \tag{6.2}$$
$$-\Delta^a(\gamma_2, \gamma_1(s))\left(T^b(\gamma_1 \circ_s \gamma_2, u(t)) + T^b(\gamma_1 \circ_s \gamma_2^{-1}, u(t))\right)$$
$$+\Delta^b(\gamma_1, \gamma_2(t))\left(T^a(\gamma_2 \circ_t \gamma_1, v(s)) + T^a(\gamma_2 \circ_t \gamma_1^{-1}, v(s))\right).$$

In the derivation of this semidirect product algebra, the Mandelstam constraints (4.2e) have been used to bring the right-hand sides into a form linear in T. The structure constants Δ are distributional and depend just on the geometry of loops and intersection points,

$$\Delta^a(\gamma, x) = \frac{1}{2}\oint_\gamma dt\, \delta^3(\gamma(t), x)\dot{\gamma}^a(t). \tag{6.3}$$

The Poisson bracket of two loop variables is non-vanishing only when an insertion of an electric field $\tilde{E}$ in a T^a-variable coincides in Σ with (the holonomy of) another loop. For gauge group $SU(N)$, the right-hand sides of the Poisson brackets can no longer be written as linear expressions in T.

By a 'quantization of the theory in the loop representation' we will mean a representation of the loop algebra (6.2) as the commutator algebra of a set of self-adjoint operators $(\hat{T}(\gamma), \hat{T}^a(\gamma, s))$ (or an appropriately smeared version of the momentum operator $\hat{T}^a(\gamma, s)$). Secondly, the Mandelstam constraints must be incorporated in the quantum theory, for example by demanding that relations like (4.2) hold also for the corresponding quantum operators $\hat{T}(\gamma)$. For the momentum Wilson loops there are analogous Mandelstam constraints.

The completion of this quantization program is different for gauge theory and gravity, in line with the remarks made in Section 2 above. For Yang–Mills theory, one has to find a Hilbert space of wave functionals that depend on either the Wilson loops $T(\gamma)$ or directly on the elements γ of some appropriate (quotient of a) loop space. The quantized Wilson loops must be self-adjoint with respect to the inner product on this Hilbert space, since they correspond to genuine observables in the classical theory. Then the Yang–Mills Hamiltonian (2.8) has to be re-expressed as a function of Wilson loops, and finally, one has to solve the eigenvector equation

$$\hat{H}_{YM}(T(\gamma), T^a(\gamma))\Psi(\alpha) = E\Psi(\alpha), \tag{6.4}$$

to obtain the spectrum of the theory. Because of the lack of an appropriate scalar product in the continuum theory, the only progress that has been made along these lines is in the lattice-regularized version of the gauge theory (see, for example, [11, 15]).

For the case of general relativity one does not require the loop variables $(T(\gamma), T^a(\gamma, s))$ to be represented self-adjointly, because they do not constitute observables on the classical phase space. At least one such representation (where the generalized Wilson loops are represented as differential operators on a space of loop functionals $\tilde{\Psi}(\gamma)$) is known, and has been extensively used in quantum gravity. The completion of the loop quantization program for gravity requires an appropriate inclusion of the 3-dimensional diffeomorphism constraints C_a and the Hamiltonian constraint C, cf. (2.6), which again means that they have to be re-expressed in terms of loop variables. Then, solutions to these quantum constraints have to be found, i.e. loop functionals that lie in their kernel,

$$\hat{C}_a(T(\gamma), T^a(\gamma))\Psi(\alpha) = 0 \ \wedge \ \hat{C}(T(\gamma), T^a(\gamma))\Psi(\alpha) = 0. \tag{6.5}$$

Finally, the space of these solutions (or an appropriate subspace) has to be given a Hilbert space structure to make a physical interpretation possible. No suitable scalar product is known at the moment, but there exist a large number of solutions to the quantum constraint equations (6.5). These solutions and their relation to knot invariants are the subject of Pullin's chapter [29].

There is one more interesting mathematical structure in the loop approach to which I would like to draw attention, and which tries to establish a relation between the quantum connection and the quantum loop representation(s). This is the so-called loop transform, first introduced by Rovelli and Smolin [30], which is supposed to intertwine the two types of representations according to

$$\tilde{\Psi}(\gamma) = \int_{\mathcal{A}/\mathcal{G}} [dA]_\mathcal{G}\, T_A(\gamma)\, \Psi_{\text{phy}}(A) \tag{6.6}$$

and

$$\left(\hat{T}(\alpha)\tilde{\Psi}\right)(\gamma) := \int_{\mathcal{A}/\mathcal{G}} [dA]_\mathcal{G}\, T_A(\gamma) \left(\hat{T}(\alpha)\Psi_{\text{phy}}(A)\right), \tag{6.7}$$

and similarly for the action of the $\hat{T}^a(\gamma)$. Indeed, it was by the heuristic use of (6.6) and (6.7) that Rovelli and Smolin arrived at the quantum representation of the loop algebra (6.2) mentioned above. The relation (6.6) is to be thought of as a non-linear analogue of the Fourier transform

$$\tilde{\Psi}(p) = \frac{1}{\sqrt{2\pi}} \int dx\, e^{ixp}\Psi(x) \tag{6.8}$$

on $L^2(\mathbf{R}, dx)$, but does in general not possess an inverse. Again, for Yang–Mills theory we do not know how to construct a concrete measure $[dA]_\mathcal{G}$ that would allow us to proceed with a loop quantization program, although in this case the loop transform can be given a rigorous mathematical meaning.

For the application to gravity, the domain space of the integration will be a smaller space $\mathcal{M}$, as explained in Section 2 above, but for this case we know even less about the well-definedness of expressions like (6.6) and (6.7). There is, however, a variety of simpler examples (such as 2+1-dimensional gravity, electrodynamics, etc.) where the loop transform is a very useful and mathematically well-defined tool.

The following diagram (Fig. 4) summarizes the relations between connection and loop dynamics, both classically and quantum-mechanically, as outlined in Sections 4 and 6.

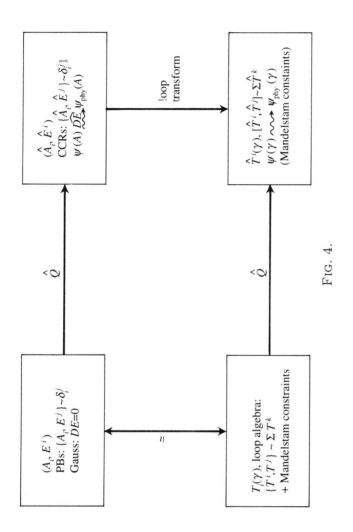

FIG. 4.

Acknowledgements

Warm thanks to John Baez for organizing an inspiring workshop.

Bibliography

1. Aref'eva, I. Ya.: The gauge field as chiral field on the path and its integrability, *Lett. Math. Phys.* **3** (1979) 241–247

2. Arms, J. M., Marsden, J. E., and Moncrief, V.: Symmetry and bifurcations of momentum mappings, *Commun. Math. Phys.* **78** (1981) 455–478

3. Ashtekar, A.: Lectures on non-perturbative gravity, World Scientific, Singapore, 1991

4. Ashtekar, A.: Mathematical problems of non-perturbative quantum general relativity, Les Houches lecture notes, preprint Syracuse SU-GP-92/11-2

5. Ashtekar, A. and Isham, C. J.: Representations of the holonomy algebras of gravity and non-Abelian gauge theories, *Classical & Quantum Gravity* **9** (1992) 1433–1467

6. Ashtekar, A. and Lewandowski, J.: Representation theory of analytic C*-algebras, this volume

7. Ashtekar, A. and Loll, R.: New loop representations for 2+1 gravity, Penn. State University preprint, in preparation

8. Baez, J.: Diffeomorphism-invariant generalized measures on the space of connections modulo gauge transformations, to appear in the Proceedings of the Conference on Quantum Topology, eds L. Crane and D. Yetter, hep-th 9305045

9. Baez, J.: Strings, loops, knots, and gauge fields, this volume

10. Bialynicki-Birula, I.: Gauge-invariant variables in the Yang–Mills theory, *Bull. Acad. Polon. Sci.* **11** (1963) 135–138

11. Brügmann, B.: The method of loops applied to lattice gauge theory, *Phys. Rev. D* **43** (1991) 566–579

12. Brylinski, J. M.: The Kaehler geometry of the space of knots in a smooth threefold, Penn. State pure mathematics report No. PM93 (1990)

13. Fulp, R. O.: The nonintegrable phase factor and gauge theory, in Proceedings of the 1990 Summer Institute on Differential Geometry, Symposia in Pure Mathematics Series

14. Fulp, R. O.: Connections on the path bundle of a principal fibre bundle, Math. preprint North Carolina State University

15. Gambini, R. and Setaro, L.: SU(2) QCD in the path representation, preprint Montevideo, April 1993

16. Gambini, R. and Trias, A.: Gauge dynamics in the C-representation,

Nucl. Phys. B **278** (1986) 436–448

17. Giles, R.: Reconstruction of gauge potentials from Wilson loops, *Phys. Rev. D* **24** (1981) 2160–2168

18. Goldberg, J. N., Lewandowski, J., and Stornaiolo, C.: Degeneracy in loop variables, *Commun. Math. Phys.* **148** (1992) 377–402

19. Loll, R.: Independent SU(2)-loop variables and the reduced configuration space of SU(2)-lattice gauge theory, *Nucl. Phys. B* **368** (1992) 121–142

20. Loll, R.: Canonical and BRST-quantization of constrained systems, in: Mathematical aspects of classical field theory, eds M. Gotay, J. Marsden, V. Moncrief, Contemporary Mathematics Series, vol. 12, 503–530

21. Loll, R.: Chromodynamics and gravity as theories on loop space, Penn. State University preprint CGPG-93/9-1, hep-th 9309056, September 1993

22. Loll, R.: Yang–Mills theory without Mandelstam constraints, *Nucl. Phys. B* **400** (1993) 126–144

23. Loll, R.: Loop variable inequalities in gravity and gauge theory, *Classical & Quantum Gravity* **10** (1993) 1471–1476

24. Mandelstam, S.: Quantum electrodynamics without potentials, *Ann. Phys. (NY)* **19** (1962) 1–24

25. Mandelstam, S.: Feynman rules for electromagnetic and Yang–Mills fields from the gauge-independent field-theoretic formalism, *Phys. Rev.* **175** (1968) 1580–1603

26. Migdal, A. A.: Loop equations and 1/N expansion, *Phys. Rep.* **102** (1983) 199–290

27. Nelson, J. E. and Regge, T.: 2+1 gravity for genus > 1, *Commun. Math. Phys.* **141** (1991) 211–223

28. Polyakov, A. M.: String representations and hidden symmetries for gauge fields, *Phys. Lett.* **82B** (1979) 247–250

29. Pullin, J.: The Gauss linking number in quantum gravity, this volume

30. Rovelli, C. and Smolin, L.: Loop space representation of quantum general relativity, *Nucl. Phys. B* **331** (1990) 80–152

31. Schäper, U.: Geometry of loop spaces, I. A Kaluza-Klein type point of view, preprint Freiburg THEP 91/3, March 1991, 41pp.

32. Wilson, K.: Confinement of quarks, *Phys. Rev. D* **10** (1974) 2445–2459

Representation Theory of Analytic Holonomy C*-algebras

Abhay Ashtekar

Center for Gravitational Physics and Geometry,
Pennsylvania State University,
University Park, Pennsylvania 16802, USA
(email: ashtekar@phys.psu.edu)

Jerzy Lewandowski

Department of Physics, University of Florida,[1]
Gainesville, Florida 32611, USA
(email: lewand@fuw.edu.pl)

Abstract

Integral calculus on the space $\mathcal{A}/\mathcal{G}$ of gauge-equivalent connections is developed. Loops, knots, links, and graphs feature prominently in this description. The framework is well suited for quantization of diffeomorphism-invariant theories of connections.

The general setting is provided by the Abelian C*-algebra of functions on $\mathcal{A}/\mathcal{G}$ generated by Wilson loops (i.e. by the traces of holonomies of connections around closed loops). The representation theory of this algebra leads to an interesting and powerful 'duality' between gauge equivalence classes of connections and certain equivalence classes of closed loops. In particular, regular measures on (a suitable completion of) $\mathcal{A}/\mathcal{G}$ are in 1–1 correspondence with certain functions of loops and diffeomorphism invariant measures correspond to (generalized) knot and link invariants. By carrying out a non-linear extension of the theory of cylindrical measures on topological vector spaces, a faithful, diffeomorphism-invariant measure is introduced. This measure can be used to define the Hilbert space of quantum states in theories of connections. The Wilson loop functionals then serve as the configuration operators in the quantum theory.

1 Introduction

The space $\mathcal{A}/\mathcal{G}$ of connections modulo gauge transformations plays an important role in gauge theories, including certain topological theories and general relativity [1]. In a typical set-up, $\mathcal{A}/\mathcal{G}$ is the classical configuration space. In the quantum theory, then, the Hilbert space of states

---

[1] On leave from: Instytut Fizyki Teoretycznej, Warsaw University ul. Hoza 69, 00–689 Warszawa, Poland

would consist of all square-integrable functions on $\mathcal{A}/\mathcal{G}$ with respect to some measure. Thus, the theory of integration over $\mathcal{A}/\mathcal{G}$ would lie at the heart of the quantization procedure. Unfortunately, since $\mathcal{A}/\mathcal{G}$ is an infinite-dimensional *non-linear* space, the well-known techniques to define integrals do not go through and, at the outset, the problem appears to be rather difficult. Even if this problem could be overcome, one would still face the issue of constructing self-adjoint operators corresponding to physically interesting observables. In a Hamiltonian approach, the Wilson loop functions provide a natural set of (manifestly gauge-invariant) configuration observables. The problem of constructing the analogous, manifestly gauge-invariant 'momentum observables' is difficult already in the classical theory: these observables would correspond to vector fields on $\mathcal{A}/\mathcal{G}$ and differential calculus on this space is not well developed.

Recently, Ashtekar and Isham [2] (hereafter referred to as A–I) developed an algebraic approach to tackle these problems. The A–I approach is in the setting of canonical quantization and is based on the ideas introduced by Gambini and Trias in the context of Yang–Mills theories [3] and by Rovelli and Smolin in the context of quantum general relativity [4]. Fix an n-manifold Σ on which the connections are to be defined and a compact Lie group G which will be the gauge group of the theory under consideration[2]. The first step is the construction of a C*-algebra of configuration observables—a sufficiently large set of gauge-invariant functions of connections on Σ. A natural strategy is to use the Wilson loop functions—the traces of holonomies of connections around closed loops on Σ—to generate this C*-algebra. Since these are configuration observables, they commute even in the quantum theory. The C*-algebra is therefore Abelian. The next step is to construct representations of this algebra. For this, the Gel'fand spectral theory provides a natural setting since any given Abelian C*-algebra with identity is naturally isomorphic with the C*-algebra of continuous functions on a compact, Hausdorff space, the Gel'fand spectrum $sp(C^\star)$ of that algebra. (As the notation suggests, $sp(C^\star)$ can be constructed directly from the given C*-algebra: its elements are homomorphisms from the given C*-algebra to the $\star$-algebra of complex numbers.) Consequently, every (continuous) cyclic representation of the A–I C*-algebra is of the following type: the carrier Hilbert space is $L^2(sp(C^\star), d\mu)$ for some regular measure $d\mu$ and the Wilson loop operators act (on square-integrable functions on $sp(C^\star)$) simply by multiplication. A–I pointed out that, since the elements of the C*-algebra are labelled by loops, there is a 1–1 correspondence between regular measures on $sp(C^\star)$

---

[2]In typical applications, Σ will be a Cauchy surface in an $(n+1)$-dimensional spacetime in the Lorentzian regime or an n-dimensional spacetime in the Euclidean regime. In the main body of this paper, n will be taken to be 3 and G will be taken to be $SU(2)$ both for concreteness and simplicity. These choices correspond to the simplest non-trivial applications of the framework to physics.

and certain functions on the space of loops. Diffeomorphism-invariant measures correspond to knot and link invariants. Thus, there is a natural interplay between connections and loops, a point which will play an important role in the present paper.

Note that, while the classical configuration space is $\mathcal{A}/\mathcal{G}$, the domain space of quantum states is $sp(C^\star)$. A–I showed that $\mathcal{A}/\mathcal{G}$ is naturally embedded in $sp(C^\star)$ and Rendall [5] showed that the embedding is in fact dense. Elements of $sp(C^\star)$ can be therefore thought of as *generalized* gauge-equivalent connections. To emphasize this point and to simplify the notation, let us denote the spectrum of the A–I C*-algebra by $\overline{\mathcal{A}/\mathcal{G}}$. The fact that the domain space of quantum states is a completion of—and hence larger than—the classical configuration space may seem surprising at first since in ordinary (i.e. non-relativistic) quantum mechanics both spaces are the same. The enlargement is, however, characteristic of quantum field theory, i.e. of systems with an infinite number of degrees of freedom. A–I further explored the structure of $\overline{\mathcal{A}/\mathcal{G}}$ but did not arrive at a complete characterization of its elements. They also indicated how one could make certain heuristic considerations of Rovelli and Smolin precise and associate 'momentum operators' with (closed) strips (i.e. $(n-1)$-dimensional ribbons) on Σ. However, without further information on the measure $d\mu$ on $\overline{\mathcal{A}/\mathcal{G}}$, one cannot decide if these operators can be made self-adjoint, or indeed if they are even densely defined. A–I did introduce certain measures with which the strip operators 'interact properly'. However, as they pointed out, these measures have support only on finite-dimensional subspaces of $\overline{\mathcal{A}/\mathcal{G}}$ and are therefore 'too simple' to be interesting in cases where the theory has an infinite number of degrees of freedom.

In broad terms, the purpose of this paper is to complete the A–I program in several directions. More specifically, we will present the following results:

1. We will obtain a complete characterization of elements of the Gel'fand spectrum. This will provide a degree of control on the domain space of quantum states which is highly desirable: since $\mathcal{A}/\mathcal{G}$ is a non-trivial, infinite-dimensional space which is not even locally compact, while $\overline{\mathcal{A}/\mathcal{G}}$ is a compact Hausdorff space, the completion procedure is quite non-trivial. Indeed, it is the lack of control on the precise content of the Gel'fand spectrum of physically interesting C*-algebras that has prevented the Gel'fand theory from playing a prominent role in quantum field theory so far.

2. We will present a faithful, diffeomorphism-invariant measure on $\overline{\mathcal{A}/\mathcal{G}}$. The theory underlying this construction suggests how one might introduce other diffeomorphism-invariant measures. Recently, Baez [6] has exploited this general strategy to introduce new measures. These developments are interesting from a strictly mathematical standpoint because the issue of existence of such measures was a subject of

controversy until recently[3]. They are also of interest from a physical viewpoint since one can, for example, use these measures to regulate physically interesting operators (such as the constraints of general relativity) and investigate their properties in a systematic fashion.

3. Together, these results enable us to define a Rovelli–Smolin 'loop transform' rigorously. This is a map from the Hilbert space $L^2(\overline{\mathcal{A}/\mathcal{G}}, d\mu)$ to the space of suitably regular functions of loops on Σ. Thus, quantum states can be represented by suitable functions of loops. This 'loop representation' is particularly well suited to diffeomorphism-invariant theories of connections, including general relativity.

We have also developed differential calculus on $\overline{\mathcal{A}/\mathcal{G}}$ and shown that the strip (i.e. momentum) operators are in fact densely defined, symmetric operators on $L^2(\overline{\mathcal{A}/\mathcal{G}}, d\mu)$; i.e. that they interact with our measure correctly. However, since the treatment of the strip operators requires a number of new ideas from physics as well as certain techniques from graph theory, these results will be presented in a separate work.

The methods we use can be summarized as follows.

First, we make the (non-linear) duality between connections and loops explicit. On the connection side, the appropriate space to consider is the Gel'fand spectrum $\overline{\mathcal{A}/\mathcal{G}}$ of the A–I C*-algebra. On the loop side, the appropriate object is the space of hoops—i.e. holonomically equivalent loops. More precisely, let us consider based, piecewise *analytic* loops and regard two as being equivalent if they give rise to the same holonomy (evaluated at the basepoint) for *any* (*G*-)connection. Call each 'holonomic equivalence class' a (*G*-)hoop. It is straightforward to verify that the set of hoops has the structure of a group. (It also carries a natural topology [7], which, however, will not play a role in this paper.) We will denote it by $\mathcal{HG}$ and call it the *hoop group*. It turns out that $\mathcal{HG}$ and the spectrum $\overline{\mathcal{A}/\mathcal{G}}$ can be regarded as being 'dual' to each other in the following sense:

1. Every element of $\mathcal{HG}$ defines a continuous, complex-valued function on $\overline{\mathcal{A}/\mathcal{G}}$ and these functions generate the algebra of all continuous functions on $\overline{\mathcal{A}/\mathcal{G}}$.

2. Every element of $\overline{\mathcal{A}/\mathcal{G}}$ defines a homomorphism from $\mathcal{HG}$ to the gauge group G (which is unique modulo an automorphism of G) and every homomorphism defines an element of $\overline{\mathcal{A}/\mathcal{G}}$.

In the case of topological vector spaces, the duality mapping respects the topology *and* the linear structure. In the present case, however, $\mathcal{HG}$ has the structure only of a group and $\overline{\mathcal{A}/\mathcal{G}}$ of a topological space. The 'duality' mappings can therefore respect only these structures. Property 2 above

---

[3] We should add, however, that in most of these debates, it was $\mathcal{A}/\mathcal{G}$—rather than its completion $\overline{\mathcal{A}/\mathcal{G}}$—that was the center of attention.

provides us with a complete (algebraic) characterization of the elements of Gel'fand spectrum $\overline{\mathcal{A}/\mathcal{G}}$ while property 1 specifies the topology of the spectrum.

The second set of techniques involves a family of projections from the topological space $\overline{\mathcal{A}/\mathcal{G}}$ to certain finite-dimensional manifolds. For each subgroup of the hoop group $\mathcal{HG}$, generated by n independent hoops, we define a projection from $\overline{\mathcal{A}/\mathcal{G}}$ to the quotient, G^n/Ad, of the nth power of the gauge group by the adjoint action. (An element of G^n/Ad is thus an equivalence class of n-tuples, $(g_1, \ldots, g_n)$, with $g_i \in G$, where $(g_1, \ldots, g_n) \sim (g_0^{-1} g_1 g_0, \ldots, g_0^{-1} g_n g_0)$ for any $g_0 \in G$.) The lifts of functions on G^n/Ad are then the non-linear analogs of the cylindrical functions on topological vector spaces. Finally, since each G^n/Ad is a compact topological space, we can equip it with measures to integrate functions in the standard fashion. This in turn enables us to define cylinder measures on $\overline{\mathcal{A}/\mathcal{G}}$ and integrate cylindrical functions thereon. Using the Haar measure on G, we then construct a natural, diffeomorphism-invariant, faithful measure on $\overline{\mathcal{A}/\mathcal{G}}$.

Finally, for completeness, we will summarize the strategy we have adopted to introduce and analyze the 'momentum operators', although, as mentioned above, these results will be presented elsewhere. The first observation is that we can define 'vector fields' on $\overline{\mathcal{A}/\mathcal{G}}$ as derivations on the ring of cylinder functions. Then, to define the specific vector fields corresponding to the momentum (or strip) operators we introduce certain notions from graph theory. These specific vector fields are 'divergence-free' with respect to our cylindrical measure on $\overline{\mathcal{A}/\mathcal{G}}$, i.e. they leave the cylindrical measure invariant. Consequently, the momentum operators are densely defined and symmetric on the Hilbert space of square-integrable functions on $\overline{\mathcal{A}/\mathcal{G}}$.

The paper is organized as follows. Section 2 is devoted to preliminaries. In Section 3, we obtain the characterization of elements of $\overline{\mathcal{A}/\mathcal{G}}$ and also present some examples of those elements which do not belong to $\mathcal{A}/\mathcal{G}$, i.e. which represent genuine, generalized (gauge equivalence classes of) connections. Using the machinery introduced in this characterization, in Section 4 we introduce the notion of cylindrical measures on $\overline{\mathcal{A}/\mathcal{G}}$ and then define a natural, faithful, diffeomorphism-invariant, cylindrical measure. Section 5 contains concluding remarks.

In all these considerations, the piecewise analyticity of loops on Σ plays an important role. That is, the specific constructions and proofs presented here would not go through if the loops were allowed to be just piecewise smooth. However, it is far from being obvious that the final results would not go through in the smooth category. To illustrate this point, in Appendix A we restrict ourselves to $U(1)$ connections and, by exploiting the Abelian character of this group, show how one can obtain the main results of this paper using piecewise C^1 loops. Whether similar constructions are possible in the non-Abelian case is, however, an open question. Finally, in Appendix

B we consider another extension. In the main body of the paper, Σ is a 3-manifold and the gauge group G is taken to be $SU(2)$. In Appendix B, we indicate the modifications required to incorporate $SU(N)$ and $U(N)$ connections on an n-manifold (keeping, however, piecewise analytic loops).

2   Preliminaries

In this section, we will introduce the basic notions, fix the notation, and recall those results from the A–I framework which we will need in the subsequent discussion.

Fix a 3-dimensional, real, analytic, connected, orientable manifold Σ. Denote by $P(\Sigma, SU(2))$ a principal fibre bundle P over the base space Σ and the structure group $SU(2)$. Since any $SU(2)$ bundle over a 3-manifold is trivial we take P to be the product bundle. Therefore, $SU(2)$ connections on P can be identified with $su(2)$-valued 1-form fields on Σ. Denote by $\mathcal{A}$ the space of smooth (say C^1) $SU(2)$ connections, equipped with one of the standard (Sobolev) topologies (see, e.g., [8]). Every $SU(2)$-valued function g on Σ defines a gauge transformation in $\mathcal{A}$,

$$A \cdot g := g^{-1}Ag + g^{-1}dg. \tag{2.1}$$

Let us denote the quotient of $\mathcal{A}$ under the action of this group $\mathcal{G}$ (of local gauge transformations) by $\mathcal{A}/\mathcal{G}$. Note that the affine space structure of $\mathcal{A}$ is lost in this projection; $\mathcal{A}/\mathcal{G}$ is a genuinely non-linear space with a rather complicated topological structure.

The next notion we need is that of closed loops in Σ. Let us begin by considering continuous, piecewise analytic (C^ω) parametrized paths, i.e. maps

$$p : [0, s_1] \cup \ldots \cup [s_{n-1}, 1] \to \Sigma \tag{2.2}$$

which are continuous on the whole domain and C^ω on the closed intervals $[s_k, s_{k+1}]$. Given two paths $p_1 : [0,1] \to \Sigma$ and $p_2 : [0,1] \to \Sigma$ such that $p_1(1) = p_2(0)$, we denote by $p_2 \circ p_1$ the natural composition:

$$p_2 \circ p_1(s) = \begin{cases} p_1(2s), & \text{for } s \in [0, \frac{1}{2}] \\ p_2(2s - 1), & \text{for } s \in [\frac{1}{2}, 1]. \end{cases} \tag{2.3}$$

The 'inverse' of a path $p: [0,1] \to \Sigma$ is a path

$$p^{-1}(s) := p(1 - s). \tag{2.4}$$

A path which starts and ends at the same point is called a *loop*. We will be interested in *based* loops. Let us therefore fix, once and for all, a point $x^0 \in \Sigma$. Denote by $\mathcal{L}_{x^0}$ the set of piecewise C^ω loops which are based at x^0, i.e. which start and end at x^0.

Given a connection A on $P(\Sigma, SU(2))$, a loop $\alpha \in \mathcal{L}_{x^0}$, and a fixed point $\hat{x}^0$ in the fiber over the point x^0, we denote the corresponding element of the holonomy group by $H(\alpha, A)$. (Since P is a product bundle, we will make the obvious choice $\hat{x}^0 = (x^0, e)$, where e is the identity in $SU(2)$.) Using the space of connections $\mathcal{A}$, we now introduce a key equivalence relation on $\mathcal{L}_{x^0}$:

Two loops $\alpha, \beta \in \mathcal{L}_{x^0}$ *will be said to be holonomically equivalent,* $\alpha \sim \beta$, *iff* $H(\alpha, A) = H(\beta, A)$, *for every* $SU(2)$ *connection A in* $\mathcal{A}$.

Each holonomic equivalence class will be called a *hoop*. Thus, for example, two loops (which have the same orientation and) which differ only in their parametrization define the same hoop. Similarly, if two loops differ just by a retraced segment, they belong to the same hoop. (More precisely, if paths p_1 and p_2 are such that $p_1(1) = p_2(0)$ and $p_1(0) = p_2(1) = x^0$, so that $p_2 \circ p_1$ is a loop in $\mathcal{L}_{x^0}$, and if ρ is a path starting at the point $p_1(1) \equiv p_2(0)$, then $p_2 \circ p_1$ and $p_2 \circ \rho \circ \rho^{-1} \circ p_1$ define the same hoop.) Note that in general the equivalence relation depends on the choice of the gauge group, whence the hoop should in fact be called an $SU(2)$-hoop. However, since the group is fixed to be $SU(2)$ in the main body of this paper, we drop this suffix. The hoop to which a loop α belongs will be denoted by $\tilde{\alpha}$.

The collection of all hoops will be denoted by $\mathcal{HG}$; thus $\mathcal{HG} = \mathcal{L}_{x^0}/\sim$. Note that, because of the properties of the holonomy map, the space of hoops, $\mathcal{HG}$, has a natural group structure: $\tilde{\alpha} \cdot \tilde{\beta} = \widetilde{\alpha \circ \beta}$. We will call it the *hoop group*[4].

With this machinery at hand, we can now introduce the A–I C*-algebra. We begin by assigning to every $\tilde{\alpha} \in \mathcal{HG}$ a real-valued function $T_{\tilde{\alpha}}$ on $\mathcal{A}/\mathcal{G}$, bounded between -1 and 1:

$$T_{\tilde{\alpha}}(\tilde{A}) := \tfrac{1}{2}\mathrm{Tr}\, H(\alpha, A), \tag{2.5}$$

where $\tilde{A} \in \mathcal{A}/\mathcal{G}$; A is any connection in the gauge equivalence class $\tilde{A}$; α any loop in the hoop $\tilde{\alpha}$; and where the trace is taken in the fundamental representation of $SU(2)$. ($T_{\tilde{\alpha}}$ is called the Wilson loop function in the physics literature.) Owing to $SU(2)$ trace identities, products of these functions can be expressed as sums:

$$T_{\tilde{\alpha}} T_{\tilde{\beta}} = \tfrac{1}{2}(T_{\tilde{\alpha} \cdot \tilde{\beta}} + T_{\tilde{\alpha} \cdot \tilde{\beta}^{-1}}), \tag{2.6}$$

where, on the right, we have used the composition of hoops in $\mathcal{HG}$. Therefore, the complex vector space spanned by the $T_{\tilde{\alpha}}$ functions is closed under the product; it has the structure of a $\star$-algebra. Denote it by $\mathcal{HA}$ and call it the *holonomy algebra*. Since each $T_{\tilde{\alpha}}$ is a bounded continuous func-

---

[4]The hoop equivalence relation seems to have been introduced independently by a number of authors in different contexts (e.g. by Gambini and collaborators in their investigation of Yang–Mills theory and by Ashtekar in the context of 2+1 gravity). Recently, the group structure of $\mathcal{HG}$ has been systematically exploited and extended by Di Bartolo, Gambini, and Griego [9] to develop a new and potentially powerful framework for gauge theories.

tion on $\mathcal{A}/\mathcal{G}$, $\mathcal{HA}$ is a subalgebra of the C*-algebra of all complex-valued, bounded, continuous functions thereon. The completion $\overline{\mathcal{HA}}$ of $\mathcal{HA}$ (under the sup norm) has therefore the structure of a C*-algebra. This is the A–I C*-algebra.

As in [2], one can make the structure of $\overline{\mathcal{HA}}$ explicit by constructing it, step by step, from the space of loops $\mathcal{L}_{x^0}$. Denote by $\mathcal{FL}_{x^0}$ the free vector space over complexes generated by $\mathcal{L}_{x^0}$. Let K be the subspace of $\mathcal{FL}_{x^0}$ defined as follows:

$$\sum_i a_i \alpha_i \in K \quad \text{iff} \quad \sum_i a_i T_{\tilde{\alpha}_i}(\tilde{A}) = 0, \quad \forall \tilde{A} \in \mathcal{A}/\mathcal{G}. \tag{2.7}$$

It is then easy to verify that $\mathcal{HA} = \mathcal{FL}_{x^0}/K$. (Note that the K-equivalence relation subsumes the hoop equivalence.) Thus, elements of $\mathcal{HA}$ can be represented either as $\sum a_i T_{\tilde{\alpha}_i}$ or as $\sum a_i[\alpha_i]_K$. The $\star$-operation, the product, and the norm on $\mathcal{HA}$ can be specified directly on $\mathcal{FL}_{x^0}/K$ as follows:

$$\sum_i (a_i\,[\alpha_i]_K)^\star = \sum_i \bar{a}_i\,[\alpha_i]_K,$$

$$\left(\sum_i a_i\,[\alpha_i]_K\right)\cdot\left(\sum_j b_j\,[\beta_j]_K\right) = \sum_{ij} a_i b_j([\alpha_i \circ \beta_j]_K + [\alpha_i \circ \beta_j^{-1}]_K),$$

$$\left\|\sum_i a_i\,[\alpha_i]_K\right\| = \sup_{\tilde{A}\in\mathcal{A}/\mathcal{G}} \left|\sum_i a_i\,T_{\tilde{\alpha}_i}(\tilde{A})\right|, \tag{2.8}$$

where $\bar{a}_i$ is the complex conjugate of a_i. The C*-algebra $\overline{\mathcal{HA}}$ is obtained by completing $\mathcal{HA}$ in the norm topology.

By construction, $\overline{\mathcal{HA}}$ is an Abelian C*-algebra. Furthermore, it is equipped with the identity element $T_{\tilde{o}} \equiv [o]_K$, where o is the trivial (i.e. zero) loop. Therefore, we can directly apply the Gel'fand representation theory. Let us summarize the main results that follow [2]. First, we know that $\overline{\mathcal{HA}}$ is naturally isomorphic to the C*-algebra of complex-valued, continuous functions on a compact Hausdorff space, the spectrum $sp(\overline{\mathcal{HA}})$. Furthermore, one can show [5] that the space $\mathcal{A}/\mathcal{G}$ of connections modulo gauge transformations is densely embedded in $sp(\overline{\mathcal{HA}})$. Therefore, we can denote the spectrum as $\overline{\mathcal{A}/\mathcal{G}}$. Second, every continuous, cyclic representation of $\overline{\mathcal{HA}}$ by bounded operators on a Hilbert space has the following form: the representation space is the Hilbert space $L^2(\overline{\mathcal{A}/\mathcal{G}}, d\mu)$ for some regular measure $d\mu$ on $\overline{\mathcal{A}/\mathcal{G}}$ and elements of $\overline{\mathcal{HA}}$, regarded as functions on $\overline{\mathcal{A}/\mathcal{G}}$, act simply by (pointwise) multiplication. Finally, each of these representations is uniquely determined (via the Gel'fand–Naimark–Segal construction) by a positive linear functional $\langle\cdot\rangle$ on the C*-algebra $\overline{\mathcal{HA}} : \sum a_i T_{\tilde{\alpha}_i} \mapsto \langle\sum a_i T_{\tilde{\alpha}_i}\rangle \equiv \sum a_i \langle T_{\tilde{\alpha}_i}\rangle$.

The functionals $\langle\cdot\rangle$ naturally define functions $\Gamma(\alpha)$ on the space $\mathcal{L}_{x^0}$ of loops—which in turn serve as the 'generating functionals' for the representation—via $\Gamma(\alpha) \equiv \langle T_{\tilde{\alpha}} \rangle$. Thus, there is a canonical correspondence between regular measures on $\overline{\mathcal{A}/\mathcal{G}}$ and generating functionals $\Gamma(\alpha)$. The question naturally arises: can we write down necessary and sufficient conditions for a function on the loop space $\mathcal{L}_{x^0}$ to qualify as a generating functional? Using the structure of the algebra $\mathcal{HA}$ outlined above, one can answer this question affirmatively:

A function $\Gamma(\alpha)$ on $\mathcal{L}_{x^0}$ serves as a generating functional for the GNS construction if and only if it satisfies the following two conditions:

$$\sum_i a_i \, T_{\tilde{\alpha}_i} = 0 \quad \Rightarrow \quad \sum_i a_i \, \Gamma(\alpha_i) = 0; \quad \text{and}$$

$$\sum_{i,j} \bar{a}_i a_j (\Gamma(\alpha_i \circ \alpha_j) + \Gamma(\alpha_i \circ \alpha_j^{-1})) \geq 0. \tag{2.9}$$

The first condition ensures that the functional $\langle \cdot \rangle$ on $\mathcal{HA}$, obtained by extending $\Gamma(\alpha)$ by linearity, is well defined while the second condition (by (2.6)) ensures that it satisfies the 'positivity' property, $\langle B^\star B \rangle \geq 0$, $\forall B \in \mathcal{HA}$. Thus, together, they provide a positive linear functional on the $\star$-algebra $\mathcal{HA}$. Finally, since $\mathcal{HA}$ is dense in $\overline{\mathcal{HA}}$ and contains the identity, it follows that the positive linear functional $\langle \cdot \rangle$ extends to the full C*-algebra $\overline{\mathcal{HA}}$.

Thus, a loop function $\Gamma(\alpha)$ satisfying (2.9) determines a cyclic representation of $\overline{\mathcal{HA}}$ and therefore a regular measure on $\overline{\mathcal{A}/\mathcal{G}}$. Conversely, every regular measure on $\overline{\mathcal{A}/\mathcal{G}}$ provides (through vacuum expectation values) a loop functional $\Gamma(\alpha) \equiv \langle T_{\tilde{\alpha}} \rangle$ satisfying (2.9). Note, finally, that the first equation in (2.9) ensures that the generating function factors through the hoop equivalence relation $\Gamma(\alpha) \equiv \Gamma(\tilde{\alpha})$.

We thus have an interesting interplay between connections and loops, or, more precisely, generalized gauge-equivalent connections (elements of $\overline{\mathcal{A}/\mathcal{G}}$) and hoops (elements of $\mathcal{HG}$). The generators $T_{\tilde{\alpha}}$ of the holonomy algebra (i.e. configuration operators) are labelled by elements $\tilde{\alpha}$ of $\mathcal{HG}$. The elements of the representation space (i.e. quantum states) are L^2 functions on $\overline{\mathcal{A}/\mathcal{G}}$. Regular measures $d\mu$ on $\overline{\mathcal{A}/\mathcal{G}}$ are, in turn, determined by functions on $\mathcal{HG}$ (satisfying (2.9)). The group of diffeomorphisms on Σ has a natural action on the algebra $\overline{\mathcal{HA}}$, its spectrum $\overline{\mathcal{A}/\mathcal{G}}$, and the space of hoops, $\mathcal{HG}$. The measure $d\mu$ is invariant under this induced action on $\overline{\mathcal{A}/\mathcal{G}}$ iff $\Gamma(\tilde{\alpha})$ is invariant under the induced action on $\mathcal{HG}$. Now, a diffeomorphism-invariant function of hoops is a function of generalized knots—generalized, because our loops are allowed to have kinks, intersections, and segments that may be traced more than once.[5] Hence, there is a 1–1 correspondence

---

[5] If the gauge group were more general, we could not have expressed products of the generators $T_{\tilde{\alpha}}$ as sums (eqn (2.6)). For $SU(3)$, for example, one cannot get rid of double products; triple (and higher) products can, however, be reduced to linear combinations

between generalized knot invariants (which in addition satisfy (2.9)) and regular diffeomorphism-invariant measures on generalized gauge-equivalent connections.

3   The spectrum

The constructions discussed in Section 2 have several appealing features. In particular, they open up an algebraic approach to the integration theory on the space of gauge-equivalent connections and bring to the forefront the duality between loops and connections. However, in practice, a good control on the precise content and structure of the Gel'fand spectrum $\overline{\mathcal{A}/\mathcal{G}}$ of $\overline{\mathcal{H}\mathcal{A}}$ is needed to make further progress. Even for simple algebras which arise in non-relativistic quantum mechanics—such as the algebra of almost periodic functions on R^n—the spectrum is rather complicated; while R^n is densely embedded in the spectrum, one does not have as much control as one would like on the points that lie in its closure (see, e.g., [5]). In the case of a C*-algebra of all continuous, bounded functions on a completely regular topological space, the spectrum is the Stone–Čech compactification of that space, whence the situation is again rather unruly. In the case of the A–I algebra $\overline{\mathcal{H}\mathcal{A}}$, the situation seems to be even more complicated: since the algebra $\mathcal{H}\mathcal{A}$ is generated only by *certain* functions on $\mathcal{A}/\mathcal{G}$, at least at first sight, $\overline{\mathcal{H}\mathcal{A}}$ appears to be a *proper* subalgebra of the C*-algebra of all continuous functions on $\mathcal{A}/\mathcal{G}$, and therefore outside the range of applicability of standard theorems.

However, the holonomy C*-algebra is also rather special: it is constructed from natural geometrical objects—connections and loops—on the 3-manifold Σ. Therefore, one might hope that its spectrum can be characterized through appropriate geometric constructions. We shall see in this section that this is indeed the case. The specific techniques we use rest heavily on the fact that the loops involved are all piecewise *analytic*.

This section is divided into three parts. In the first, we introduce the key tools that are needed (also in subsequent sections), in the second, we present the main result, and in the third we give examples of 'generalized gauge-equivalent connections', i.e. of elements of $\overline{\mathcal{A}/\mathcal{G}} - \mathcal{A}/\mathcal{G}$. To keep the discussion simple, generally we will not explicitly distinguish between paths and loops and their images in Σ.

3.1   Loop decomposition

A key technique that we will use in various constructions is the decomposition of a finite number of hoops, $\tilde{\alpha}_1, \ldots, \tilde{\alpha}_k$, into a finite number of *independent* hoops, $\tilde{\beta}_1, \ldots, \tilde{\beta}_n$. The main point here is that a given set

---

of double products of $T_{\tilde{\alpha}}$ and the $T_{\tilde{\alpha}}$ themselves. In this case, the generating functional is defined on single *and double* loops, whence, in addition to knot invariants, link invariants can also feature in the definition of the generating function.

of loops need not be 'holonomically independent': every open segment in a given loop may be shared by other loops in the given set, whence, for *any* connection A in $\mathcal{A}$, the holonomies around these loops could be inter-related. However, using the fact that the loops are piecewise analytic, we will be able to show that any finite set of loops *can* be decomposed into a finite number of independent segments and this in turn leads us to the desired independent hoops. The availability of this decomposition will be used in the next subsection to obtain a characterization of elements of $\overline{\mathcal{A}/\mathcal{G}}$ and in Section 4 to define 'cylindrical' functions on $\overline{\mathcal{A}/\mathcal{G}}$.

Our aim then is to show that, given a finite number of hoops, $\tilde{\alpha}_1, \ldots, \tilde{\alpha}_k$, (with $\tilde{\alpha}_i \neq$ identity in $\mathcal{HG}$ for any i), there exists a finite set of loops, $\{\beta_1, \ldots, \beta_n\} \subset \mathcal{L}_{x^0}$, which satisfies the following properties:

1. If we denote by $\mathcal{HG}(\tilde{\gamma}_1, \ldots, \tilde{\gamma}_m)$ the subgroup of the hoop group $\mathcal{HG}$ generated by the hoops $\tilde{\gamma}_1, \ldots, \tilde{\gamma}_m$, then we have

$$\mathcal{HG}(\tilde{\alpha}_1, \ldots, \tilde{\alpha}_k) \subset \mathcal{HG}(\tilde{\beta}_1, \ldots, \tilde{\beta}_n), \tag{3.1}$$

   where, as before, $\tilde{\beta}_i$ denotes the hoop to which β_i belongs.

2. Each of the new loops β_i contains an open segment (i.e. an embedding of an interval) which is traced exactly once and which intersects any of the remaining loops β_j (with $i \neq j$) at most at a finite number of points.

The first condition makes the sense in which each hoop $\tilde{\alpha}_i$ can be decomposed in terms of $\tilde{\beta}_j$ precise while the second condition specifies the sense in which the new hoops $\tilde{\beta}_1, \ldots, \tilde{\beta}_n$ are independent.

Let us begin by choosing a loop $\alpha_i \in \mathcal{L}_{x^0}$ in each hoop $\tilde{\alpha}_i$, for all $i = 1, \ldots, k$, such that none of the $\tilde{\alpha}_i$ has a piece which is immediately retraced (i.e. none of the loops contains a path of the type $\rho \cdot \rho^{-1}$). Now, *since the loops are piecewise analytic, any two which overlap do so either on a finite number of finite intervals and/or intersect in a finite number of points.* Ignore the isolated intersection points and mark the end points of all the overlapping intervals (including the end points of self-overlaps of a loop). Let us call these marked points *vertices*. Next, divide each loop α_i into paths which join consecutive vertices. Thus, each of these paths is piecewise analytic, is part of at least one of the loops $\alpha_1, \ldots, \alpha_k$, and has a nonzero (parameter) measure along any loop to which it belongs. Two distinct paths intersect at a finite set of points. Let us call these (oriented) paths *edges*. Denote by n the number of edges that result. The edges and vertices form a (piecewise analytically embedded) graph. By construction, each loop in the list is contained in the graph and, ignoring parametrization, can be expressed as a composition of a finite number of its n edges. Finally, this decomposition is 'minimal' in the sense that, if the initial set $\alpha_1, \ldots, \alpha_k$ of loops is extended to include p additional loops, $\alpha_{k+1}, \ldots, \alpha_{k+p}$, each edge in the original decomposition of k loops is expressible as a product of the edges which feature in the decomposition of the $k + p$ loops.

The next step is to convert this edge decomposition of loops into a decomposition in terms of elementary loops. This is easy to achieve. Connect each vertex v to the basepoint x^0 by an oriented, piecewise analytic curve $q(v)$ such that these curves overlap with any edge e_i *at most* at a finite number of isolated points. Consider the closed curves β_i, starting and ending at the basepoint x^0,

$$\beta_i := q(v_i^+) \circ e_i \circ (q(v_i^-))^{-1}, \qquad (3.2)$$

where $v_i^\pm$ are the end points of the edge e_i, and denote by $\mathcal{S}$ the set of these n loops. Loops β_i are *not* unique because of the freedom in choosing the curves $q(v)$. However, we will show that they provide us with the required set of independent loops associated with the given set $\{\alpha_1, \ldots, \alpha_k\}$.

Let us first note that the decomposition satisfies certain conditions which follow immediately from the properties of the segments noted above:

1. $\mathcal{S}$ is a finite set and every $\beta_i \in \mathcal{S}$ is a non-trivial loop in the sense that the hoop $\tilde{\beta}_i$ it defines is not the identity element of $\mathcal{HG}$;

2. every loop β_i contains an open interval which is traversed exactly once and no finite segment of which is shared by any other loop in $\mathcal{S}$;

3. every hoop $\tilde{\alpha}_1, \ldots, \tilde{\alpha}_k$ in our initial set can be expressed as a finite composition of hoops defined by elements of $\mathcal{S}$ and their inverses (where a loop β_i and its inverse β_i^{-1} may occur more than once); and,

4. if the initial set $\alpha_1, \ldots, \alpha_k$ of loops is extended to include p additional loops, $\alpha_{k+1}, \ldots, \alpha_{k+p}$, and if $\mathcal{S}'$ is the set of n' loops $\beta_1', \ldots, \beta_{n'}'$ corresponding to $\alpha_1, \ldots, \alpha_{k+p}$, such that the paths $q'(v')$ agree with the paths $q(v)$ whenever $v = v'$, then each hoop $\tilde{\beta}_i$ is a finite product of the hoops $\tilde{\beta}_j'$.

These properties will play an important role throughout the paper. For the moment we only note that we have achieved our goal: the property 2 above ensures that the set $\{\beta_1, \ldots, \beta_n\}$ of loops is independent in the sense we specified in the point 2 (just below eqn (3.1)) and that 3 above implies that the hoop group generated by $\alpha_1, \ldots, \alpha_k$ is contained in the hoop group generated by $\beta_1, \ldots, \beta_n$, i.e. that our decomposition satisfies (3.1).

We will conclude this subsection by showing that the independence of hoops $\{\tilde{\beta}_1, \ldots, \tilde{\beta}_n\}$ has an interesting consequence which will be used repeatedly:

Lemma 3.1 *For every* $(g_1, \ldots, g_n) \in [SU(2)]^n$, *there exists a connection* $A_0 \in \mathcal{A}$ *such that*

$$H(\tilde{\beta}_i, A_0) = g_i \quad \forall i = 1, \ldots, n. \qquad (3.3)$$

Proof Fix a sequence of elements $(g_1, \ldots, g_n)$ of $SU(2)$, i.e. a point in $[SU(2)]^n$. Next, for each loop β_i, pick an Abelian connection $A'_{(i)} = a'_{(i)} w'$ where $a'_{(i)}$ is a real-valued 1-form on Σ and w' a fixed element of the Lie algebra of $SU(2)$. Let furthermore the support of $a'_{(i)}$ intersect the ith segment s_i in a finite interval, and, if $j \neq i$, intersect the jth segment s_j only on a set of measure zero (as measured along any loop α_m containing that segment). Finally, let us suppose that the connection is 'generic' in the sense that the holonomy that the holonomy $H(\beta_i, A'_{(i)})$ is not the identity element of $SU(2)$. Then, it is of the form:

$$H(\beta_i, A'_{(i)}) = \exp w, \quad w \in su(2), \quad w \neq 0$$

with w a multiple of w'. Now, consider a connection

$$A_{(i)} = t\, g^{-1} \cdot A'_{(i)} \cdot g, \quad g \in SU(2), \quad t \in \mathbf{R}$$

Then, we have

$$H(\beta_i, A_{(i)}) = \exp t(g^{-1} w g).$$

Hence, by choosing g and t appropriately, we can make $H(\beta_i, A_{(i)})$ coincide with any element of $SU(2)$, in particular g_i. Because of the independence of the loops β_i noted above, we can choose connections $A_{(i)}$ independently for every β_i. Then, the connection

$$A_0 := A_{(1)} + \cdots + A_{(n)} \tag{3.4}$$

satisfies (3.3). □

We conclude this subsection with two remarks.

1. The result (3.3) we just proved can be taken as a statement of the independence of the $(SU(2))$-hoops $\tilde{\beta}_1, \ldots, \tilde{\beta}_n$; it captures the idea that the loops $\beta_1, \ldots, \beta_n$ we constructed above are holonomically independent as far as $SU(2)$ connections are concerned. This notion of independence is, however, weaker than the one contained in the statement 2 above. Indeed, that notion does not refer to connections at all and is therefore not tied to any specific gauge group. More precisely, we have the following. We just showed that if loops are independent in the sense of 2, they are necessarily independent in the sense of (3.3). The converse is not true. Let α and γ be two loops in $\mathcal{L}_{x^0}$ which do not share any point other than x^0 and which have no self-intersections or self-overlaps. Set $\beta_1 = \alpha$ and $\beta_2 = \alpha \cdot \gamma$. Then, it is easy to check that $\tilde{\beta}_i$, $i = 1, 2$ are independent in the sense of (3.3) although they are obviously not independent in the sense of 2. Similarly, given α, we can set $\beta = \alpha^2$. Then, $\{\beta\}$ is independent in the sense of (3.3) (since the square root is well defined in $SU(2)$) but not in the sense of 2. From now on, we will say that loops satisfying

2 are *independent* and those satisfying (3.3) are *weakly independent*. This definition extends naturally to hoops. Although we will not need them explicitly, it is instructive to spell out some algebraic consequences of these two notions. Algebraically, weak independence of hoops $\tilde{\beta}_i$ ensures that $\tilde{\beta}_i$ freely generate a subgroup of $\mathcal{HG}$. On the other hand, if $\tilde{\beta}_i$ are independent, then every homomorphism from $\mathcal{HG}(\tilde{\beta}_1, \ldots, \tilde{\beta}_n)$ to *any* group G can be extended to every finitely generated subgroup of $\mathcal{HG}$ which contains $\mathcal{HG}(\tilde{\beta}_1, \ldots, \tilde{\beta}_n)$. Weak independence will suffice for all considerations in this section. However, to ensure that the measure introduced in Section 4 is well defined, we will need hoops which are independent.

2. The availability of the loop decomposition sheds considerable light on the hoop equivalence: one can show that two loops in $\mathcal{L}_{x^0}$ define the same $(SU(2))$-hoop if and only if they differ by a combination of reparametrization and (immediate) retracings. This second characterization is useful because it involves only the loops; no mention is made of connections and holonomies. This characterization continues to hold if the gauge group is replaced by $SU(N)$ or indeed any group which has the property that, for every non-negative integer n, the group contains a subgroup which is freely generated by n elements.

3.2 Characterization of $\overline{\mathcal{A}/\mathcal{G}}$

We will now use the availability of this hoop decomposition to obtain a complete characterization of all elements of the spectrum $\overline{\mathcal{A}/\mathcal{G}}$.

The characterization is based on results due to Giles [10]. Recall first that elements of the Gel'fand spectrum $\overline{\mathcal{A}/\mathcal{G}}$ are in 1–1 correspondence with (multiplicative, $\star$-preserving) homomorphisms from $\overline{\mathcal{HA}}$ to the $\star$-algebra of complex numbers. Thus, in particular, every $\bar{A} \in \overline{\mathcal{A}/\mathcal{G}}$ acts on $T_{\tilde{\alpha}} \in \overline{\mathcal{HA}}$ and hence on any hoop $\tilde{\alpha}$ to produce a complex number $\bar{A}(\tilde{\alpha})$. Now, using certain constructions due to Giles, A–I [2] were able to show that:

Lemma 3.2 *Every element $\bar{A}$ of $\overline{\mathcal{A}/\mathcal{G}}$ defines a homomorphism $\hat{H}_{\bar{A}}$ from the hoop group $\mathcal{HG}$ to $SU(2)$ such that, $\forall \tilde{\alpha} \in \mathcal{HG}$,*

$$\bar{A}(\tilde{\alpha}) = \tfrac{1}{2} \operatorname{Tr} \hat{H}_{\bar{A}}(\tilde{\alpha}). \qquad (3.5a)$$

They also exhibited the homomorphism explicitly. (Here, we have paraphrased the A–I result somewhat because they did not use the notion of hoops. However, the step involved is completely straightforward.)

Our aim now is to establish the converse of this result. Let us make a small digression to illustrate why the converse cannot be entirely straightforward. Given a homomorphism $\hat{H}$ from $\mathcal{HG}$ to $SU(2)$ we can simply define $\bar{A}$ via $\bar{A}(\tilde{\alpha}) = \tfrac{1}{2}\operatorname{Tr} H(\tilde{\alpha})$. The question is whether this $\bar{A}$ can qualify as an element of the spectrum. From the definition, it is clear that

$$|\bar{A}(\tilde{\alpha})| \leq \|T_{\tilde{\alpha}}\| \equiv \sup_{\tilde{A} \in \mathcal{A}/\mathcal{G}} |T_{\tilde{\alpha}}(\tilde{A})| \qquad (3.6a)$$

for all hoops $\tilde{\alpha}$. On the other hand, to qualify as an element of the spectrum $\overline{\mathcal{A}/\mathcal{G}}$, the homomorphism $\bar{A}$ must be continuous, i.e. should satisfy

$$|\bar{A}(f)| \leq \|f\| \qquad (3.6b)$$

for every f $(\equiv \sum a_i [\tilde{\alpha}_i]_K) \in \mathcal{HA}$. Since, a priori, (3.6b) appears to be a stronger requirement than (3.6a), one would expect that there are more homomorphisms $\hat{H}$ from the hoop group $\mathcal{HG}$ to $SU(2)$ than the $\hat{H}_{\bar{A}}$, which arise from elements of $\overline{\mathcal{A}/\mathcal{G}}$. That is, one might expect that the converse of the A–I result would not be true. It turns out, however, that if the holonomy algebra is constructed from piecewise *analytic* loops, the apparently weaker condition (3.6a) is in fact equivalent to (3.6b), and the converse does hold. Whether it would continue to hold if one used piecewise smooth loops—as was the case in the A–I analysis—is not clear (see Appendix A).

We begin our detailed analysis with an immediate application of Lemma 3.1:

Lemma 3.3 *For every homomorphism $\hat{H}$ from $\mathcal{HG}$ to $SU(2)$, and every finite set of hoops $\{\tilde{\alpha}_1, \ldots, \tilde{\alpha}_k\}$, there exists an $SU(2)$ connection A_0 such that for every $\tilde{\alpha}_i$ in the set,*

$$\hat{H}(\tilde{\alpha}_i) = H(\alpha_i, A_0), \qquad (3.7)$$

where, as before, $H(\alpha_i, A_0)$ is the holonomy of the connection A_0 around any loop α in the hoop class $\tilde{\alpha}$.

Proof Let $(\beta_1, \ldots, \beta_n)$ be, as in Section 3.1, a set of independent loops corresponding to the given set of hoops. Denote by $(g_1, \ldots, g_n)$ the image of $(\tilde{\beta}_1, \ldots, \tilde{\beta}_n)$ under $\hat{H}$. Use the construction of Section 3.1 to obtain a connection A_0 such that $g_i = H(\tilde{\beta}_i, A_0)$ for all i. It then follows (from the definition of the hoop group) that for *any* hoop $\tilde{\gamma}$ in the subgroup $\mathcal{HG}(\tilde{\beta}_1, \ldots, \tilde{\beta}_n)$ generated by $\tilde{\beta}_1, \ldots, \tilde{\beta}_n$, we have $\hat{H}(\tilde{\gamma}) = H(\tilde{\gamma}, A_0)$. Since each of the given $\tilde{\alpha}_i$ belongs to $\mathcal{HG}(\tilde{\beta}_1, \ldots, \tilde{\beta}_n)$, we have, in particular, (3.7). $\square$

We now use this lemma to prove the main result. Recall that each $\bar{A} \in \overline{\mathcal{A}/\mathcal{G}}$ is a homomorphism from $\overline{\mathcal{HA}}$ to complex numbers and as before denote by $\bar{A}(\tilde{\alpha})$ the number assigned to $T_{\tilde{\alpha}} \in \overline{\mathcal{HA}}$ by $\bar{A}$. Then, we have:

Lemma 3.4 *Given a homomorphism $\hat{H}$ from the hoop group $\mathcal{HG}$ to $SU(2)$ there exists an element $\bar{A}_{\hat{H}}$ of the spectrum $\overline{\mathcal{A}/\mathcal{G}}$ such that*

$$\bar{A}_{\hat{H}}(\tilde{\alpha}) = \tfrac{1}{2} \mathrm{Tr} \; \hat{H}(\tilde{\alpha}). \qquad (3.5b)$$

Two homomorphisms $\hat{H}$ and $\hat{H}'$ define the same element of the spectrum *if and only if $\hat{H}'(\tilde{\alpha}) = g^{-1} \cdot \hat{H}(\tilde{\alpha}) \cdot g$, $\forall \tilde{\alpha} \in \mathcal{HG}$, for some ($\tilde{\alpha}$-independent) g in $SU(2)$.*

Proof The idea is to *define* the required $\tilde{A}_{\hat{H}}$ using (3.5b), i.e. to show that the right side of (3.5b) provides a homomorphism from $\overline{\mathcal{HA}}$ to the $\star$-algebra of complex numbers. Let us begin with the free complex vector space $\mathcal{FHG}$ over the hoop group $\mathcal{HG}$ and define on it a complex-valued function h as follows: $h(\sum a_i \tilde{\alpha}_i) := \frac{1}{2} \sum a_i \operatorname{Tr} \hat{H}(\tilde{\alpha}_i)$. We will first show that h passes through the K-equivalence relation of A–I (noted in Section 2) and is therefore a well-defined complex-valued function on the holonomy $\star$-algebra $\mathcal{HA} = \mathcal{FHG}/K$. Note first that Lemma 3.3 immediately implies that, given a finite set of hoops, $\{\tilde{\alpha}_1, \dots, \tilde{\alpha}_k\}$, there exists a connection A_0 such that

$$\left| h\left(\sum_{i=1}^{k} a_i \tilde{\alpha}_i \right) \right| = \left| \sum_{i=1}^{k} a_i T_{\tilde{\alpha}_i}(\tilde{A}_0) \right| \leq \sup_{\tilde{A} \in \mathcal{A}/\mathcal{G}} \left| \sum_{i=1}^{k} a_i T_{\tilde{\alpha}_i}(\tilde{A}) \right|. \tag{3.8}$$

Therefore, we have

$$\sum_{i=1}^{k} a_i T_{\tilde{\alpha}_i}(\tilde{A}) = 0 \ \forall \tilde{A} \in \mathcal{A}/\mathcal{G} \quad \Rightarrow \quad h\left(\sum_{i=1}^{k} a_i \tilde{\alpha}_i \right) = 0.$$

Since the left side of this implication defines the K-equivalence, it follows that the function h on $\mathcal{FHG}$ has a well-defined projection to the holonomy $\star$-algebra $\mathcal{HA}$. That h is linear is obvious from its definition. That it respects the $\star$-operation and is a multiplicative homomorphism follows from the definitions (2.8) of these operations on $\mathcal{HA}$ (and the definition of the hoop group $\mathcal{HG}$). Finally, the continuity of this funtion on $\mathcal{HA}$ is obvious from (3.8). Hence h extends uniquely to a homomorphism from the C*-algebra $\overline{\mathcal{HA}}$ to the $\star$-algebra of complex numbers, i.e. defines a unique element $\tilde{A}_{\hat{H}}$ via $\tilde{A}_{\hat{H}}(B) = h(B)$, $\forall B \in \overline{\mathcal{HA}}$.

Next, suppose $\hat{H}_1$ and $\hat{H}_2$ give rise to the same element of the spectrum. In particular, then, $\operatorname{Tr} \hat{H}_1(\tilde{\alpha}) = \operatorname{Tr} \hat{H}_2(\tilde{\alpha})$ for all hoops. Now, there is a general result: given two homomorphisms $\hat{H}_1$ and $\hat{H}_2$ from a group $\mathbf{G}$ to $SU(2)$ such that $\operatorname{Tr} \hat{H}_1(\mathbf{g}) = \operatorname{Tr} \hat{H}_2(\mathbf{g})$, $\forall \mathbf{g} \in \mathbf{G}$, there exists $g_0 \in SU(2)$ such that $\hat{H}_2(\mathbf{g}) = g_0^{-1} \cdot \hat{H}_1(\mathbf{g}) \cdot g_0$, where g_0 is independent of $\mathbf{g}$. Using the hoop group $\mathcal{HG}$ for $\mathbf{G}$, we obtain the desired uniqueness result. $\square$

To summarize, by combining the results of the three lemmas, we obtain the following characterization of the points of the Gel'fand spectrum of the holonomy C*-algebra:

Theorem 3.5 *Every point $\bar{A}$ of $\overline{\mathcal{A}/\mathcal{G}}$ gives rise to a homomorphism $\hat{H}$ from the hoop group $\mathcal{HG}$ to $SU(2)$ and every homomorphism $\hat{H}$ defines a point of $\overline{\mathcal{A}/\mathcal{G}}$, such that $\bar{A}(\tilde{\alpha}) = \frac{1}{2} \operatorname{Tr} \hat{H}(\tilde{\alpha})$. This is a 1–1 correspondence modulo the trivial ambiguity that homomorphisms $\hat{H}$ and $g^{-1} \cdot \hat{H} \cdot g$ define* *the same point of $\overline{\mathcal{A}/\mathcal{G}}$.*

We conclude this subsection with some remarks:

1. It is striking that Theorem 3.5 *does not require the homomorphism $\hat{H}$ to be continuous*; indeed, no reference is made to the topology of the hoop group anywhere. This purely algebraic characterization of the elements of $\overline{\mathcal{A}/\mathcal{G}}$ makes it convenient to use it in practice.

2. Note that the homomorphism $\hat{H}$ determines an element of $\overline{\mathcal{A}/\mathcal{G}}$ and *not* of $\mathcal{A}/\mathcal{G}$; $\bar{A}_{\hat{H}}$ is *not*, in general, a smooth (gauge-equivalent) connection. Nonetheless, as Lemma 3.3 tells us, it can be approximated by smooth connections in the sense that, given any *finite* number of hoops, one can construct a smooth connection which is indistinguishable from $\bar{A}_{\hat{H}}$ as far as these hoops are concerned. (This is stronger than the statement that $\mathcal{A}/\mathcal{G}$ is dense in $\overline{\mathcal{A}/\mathcal{G}}$.) Necessary and sufficient conditions for $\hat{H}$ to arise from a smooth connection were given by Barrett [11] (see also [7]).

3. There are several folk theorems in the literature to the effect that given a function on the loop space $\mathcal{L}_{x^0}$ satisfying certain conditions, one can reconstruct a connection (modulo gauge) such that the given function is the trace of the holonomy of that connection. (For a summary and references, see, e.g., [4].) Results obtained in this subsection have a similar flavor. However, there is a *key* difference: our main result shows the existence of a *generalized* connection (modulo gauge), i.e. an element of $\overline{\mathcal{A}/\mathcal{G}}$ rather than a regular connection in $\mathcal{A}/\mathcal{G}$. A generalized connection can also be given a geometrical meaning in terms of parallel transport, but in a *generalized* bundle [7].

3.3 Examples of $\bar{A}$

Fix a connection $A \in \mathcal{A}$. Then, the holonomy $H(\alpha, A)$ defines a homomorphism $\hat{H}_A : \mathcal{HG} \mapsto SU(2)$. A gauge-equivalent connection, $A' = g^{-1}Ag + g^{-1}dg$, gives rise to the homomorphism $\hat{H}_{A'} = g^{-1}(x^0)\hat{H}_A \, g(x^0)$. Therefore, by Theorem 3.5, A and A' define the same element of the Gel'fand spectrum; $\mathcal{A}/\mathcal{G}$ is naturally embedded in $\overline{\mathcal{A}/\mathcal{G}}$. Furthermore, Lemma 3.3 now implies that the embedding is in fact dense. This provides an alternative proof of the A–I and Rendall[6] results quoted in Section 1. Had the gauge group been different and/or had Σ a dimension greater than 3, there would exist non-trivial G-principal bundles over Σ. In this case, even if we begin with a specific bundle in our construction of the holonomy C*-algebra, connections on *all* possible bundles belong to the Gel'fand spectrum (see Appendix B). This is one illustration of the non-triviality of

---

[6]Note, however, that while this proof is tailored to the holonomy C*-algebra $\overline{\mathcal{HA}}$, Rendall's [5] proof is applicable in more general contexts.

the completion procedure leading from $\mathcal{A}/\mathcal{G}$ to $\overline{\mathcal{A}/\mathcal{G}}$.

In this subsection, we will illustrate another facet of this non-triviality: we will give examples of elements of $\overline{\mathcal{A}/\mathcal{G}}$ which do not arise from $\mathcal{A}/\mathcal{G}$. These are the *generalized* gauge-equivalent connections $\bar{A}$ which have a 'distributional character' in the sense that their support is restricted to sets of measure zero in Σ.

Recall first that the holonomy of the zero connection around any hoop is identity. Hence, the homomorphism $\hat{H}_0$ from $\mathcal{HG}$ to $SU(2)$ it defines sends every hoop $\tilde{\alpha}$ to the identity element e of $SU(2)$ and the multiplicative homomorphism it defines from the C*-algebra $\overline{\mathcal{HA}}$ to complex numbers sends each $T_{\tilde{\alpha}}$ to 1 $(= (1/2)\mathrm{Tr}\ e)$. We now use this information to introduce the notion of the support of a generalized connection. We will say that $\bar{A} \in \overline{\mathcal{A}/\mathcal{G}}$ has support on a subset S of Σ if for every hoop $\tilde{\alpha}$ which has at least one representative loop α which fails to pass through S, we have: (i) $\hat{H}_{\bar{A}}(\tilde{\alpha}) = e$, the identity element of $SU(2)$; or, equivalently, (ii) $\bar{A}([\tilde{\alpha}]_K) = 1$. (In (ii), $\bar{A}$ is regarded as a homomorphism of $\star$-algebras. While $\hat{H}_{\bar{A}}$ has an ambiguity, noted in Theorem 3.5, conditions (i) and (ii) are insensitive to it.)

We are now ready to give the simplest example of a generalized connection $\bar{A}$ which has support at a single point $x \in \Sigma$. Note first that, owing to piecewise analyticity, if $\alpha \in \mathcal{L}_{x^0}$ passes through x, it does so only a finite number of times, say N. Let us denote the incoming and outgoing tangent directions to the curve at its jth passage through x by (v_j^-, v_j^+). The generalized connections we want to define now will depend only on these tangent directions at x. Let ϕ be a (not necessarily continuous) mapping from the 2-sphere S^2 of directions in the tangent space at x to $SU(2)$. Set

$$\Phi_j = \phi(-v_j^-)^{-1} \cdot \phi(v_j^+),$$

so that, if at the jth passage, α arrives at x and then simply returns by retracing its path in a neighborhood of x, we have $\Phi_j = e$, the identity element of $SU(2)$. Now, define $\hat{H}_\phi : \mathcal{HG} \mapsto SU(2)$ via

$$\hat{H}_\phi(\tilde{\alpha}) := \begin{cases} e, & \text{if } \tilde{\alpha} \notin \mathcal{HG}_x, \\ \Phi_N \ \ldots \ \Phi_1, & \text{otherwise,} \end{cases} \tag{3.9}$$

where $\mathcal{HG}_x$ is the space of hoops every loop in which passes through x. For each choice of ϕ, the mapping $\hat{H}_\phi$ is well defined, i.e. is independent of the choice of the loop α in the hoop class $\tilde{\alpha}$ used in its construction. (In particular, we used tangent directions rather than vectors to ensure this.) It defines a homomorphism from $\mathcal{HG}$ to $SU(2)$ which is non-trivial only on $\mathcal{HG}_x$, and hence a generalized connection with support at x. Furthermore, one can verify that (3.9) is the most general element of $\overline{\mathcal{A}/\mathcal{G}}$ which has support at x and depends only on the tangent vectors at that point. Using analyticity of the loops, we can similarly construct elements which depend on the higher-order behavior of loops at x. (There is no generalized connection with support at x which depends only on the zeroth-order behavior of the curve, i.e. only on whether the curve passes through x or not.)

One can proceed in an analogous manner to produce generalized connections which have support on 1- or 2-dimensional submanifolds of Σ. We conclude with an example. Fix a 2-dimensional, oriented, analytic sub-manifold M of Σ and an element g of $SU(2)$, $(g \neq e)$. Denote by $\mathcal{HG}_M$ the subset of $\mathcal{HG}$ consisting of hoops, each loop of which intersects M (non-tangentially, i.e. such that at least one of the incoming or the outgoing tangent direction to the curve is transverse to M). As before, owing to analyticity, any $\alpha \in \mathcal{L}_{x^0}$ can intersect M only a finite number of times. Denote as before the incoming and the outgoing tangent directions at the jth intersection by v_j^- and v_j^+ respectively. Let $\epsilon(\alpha_j^\pm)$ equal 1 if the orientation of $(v_j^\pm, M)$ coincides with the orientation of Σ; -1 if the two orientations are opposite; and 0 if $v_j^\pm$ is tangential to M. Define $\hat{H}_{(g,M)} : \mathcal{HG} \mapsto SU(2)$ via

$$\hat{H}_{(g,M)}(\tilde{\alpha}) := \begin{cases} e, & \text{if } \tilde{\alpha} \notin \mathcal{HG}_M, \\ g^{\epsilon(\alpha_N^+)} \cdot g^{\epsilon(\alpha_N^-)} \quad \cdots \quad g^{\epsilon(\alpha_1^+)} \cdot g^{\epsilon(\alpha_1^-)}, & \text{otherwise,} \end{cases}$$
(3.10)

where $g^0 \equiv e$, the identity element of $SU(2)$. Again, one can check that (3.10) is a well-defined homomorphism from $\mathcal{HG}$ to $SU(2)$, which is non-trivial only on $\mathcal{HG}_M$ and therefore defines a generalized connection with support on M. One can construct more sophisticated examples in which the homomorphism is sensitive to the higher-order behavior of the loop at the points of intersection and/or the fixed $SU(2)$ element g is replaced by $g(x) : M \mapsto SU(2)$.

4   Integration on $\overline{\mathcal{A}/\mathcal{G}}$

In this section we shall develop a general strategy to perform integrals on $\overline{\mathcal{A}/\mathcal{G}}$ and introduce a natural, faithful, diffeomorphism-invariant measure on this space. The existence of this measure provides considerable confidence in the ideas involving the loop transform and the loop representation introduced in [3, 4].

The basic strategy is to carry out an appropriate generalization of the theory of cylindrical measures [12, 13] on topological vector spaces[7]. The key idea in our approach is to replace the standard linear duality on vector spaces by the duality between connections and loops. In the first part of

---

[7]This idea was developed also by Baez [6] using, however, an approach which is somewhat different from the one presented here. Chronologically, the authors of this paper first found the faithful measure introduced in this section and reported this result in the first draft of this paper. Subsequently, Baez and the authors independently realized that the theory of cylindrical measures provides the appropriate conceptual setting for this discussion.

this section, we will define cylindrical functions, and in the second part, we will present a natural cylindrical measure.

4.1   Cylindrical functions

Let us begin by recalling the situation in the linear case. Let V denote a topological vector space and V^* its dual. Given a *finite-dimensional* subspace S^* of V^*, we can define an equivalence relation on V: $v_1 \sim v_2$ if and only if $\langle v_1, s^* \rangle = \langle v_2, s^* \rangle$ for all $v_1, v_2 \in V$ and $s^* \in S^*$, where $\langle v, s^* \rangle$ is the action of $s^* \in S^*$ on $v \in V$. Denote the quotient $V/\sim$ by S. Clearly, S is a finite-dimensional vector space and we have a natural projection map $\pi(S^*) : V \mapsto S$. A function f on V is said to be *cylindrical* if it is a pullback via $\pi(S^*)$ to V of a function $\tilde{f}$ on S, for *some* choice of S^*. These are the functions we will be able to integrate. Equip each finite-dimensional vector space S with a measure $d\tilde{\mu}$ and set

$$\int_V d\mu \, f := \int_S d\tilde{\mu} \, \tilde{f}. \tag{4.1}$$

For this definition to be meaningful, however, the set of measures $d\tilde{\mu}$ that we chose on vector spaces S must satisfy certain compatibility conditions. Thus, if a function f on V is expressible as a pullback via $\pi(S_j^*)$ of functions $\tilde{f}_j$ on S_j, for $j = 1, 2$, say, we must have

$$\int_{S_1} d\tilde{\mu}_1 \, \tilde{f}_1 = \int_{S_2} d\tilde{\mu}_2 \, \tilde{f}_2. \tag{4.2}$$

Such compatible measures do exist. Perhaps the most familiar example is the (normalized) Gaussian measure.

We want to extend these ideas, using $\overline{\mathcal{A}/\mathcal{G}}$ in place of V. An immediate problem is that $\overline{\mathcal{A}/\mathcal{G}}$ is a genuinely non-linear space whence there is no obvious analog of V^* and S^* which are central to the above construction of cylindrical measures. The idea is to use the results of Section 3 to find suitable substitutes for these spaces. The appropriate choices turn out to be the following: we will let the hoop group $\mathcal{HG}$ play the role of V^* and its subgroups, generated by a finite number of independent hoops, the role of S^*. (Recall that S^* is generated by a finite number of linearly independent basis vectors.)

More precisely, we proceed as follows. From Theorem 3.5, we know that each $\bar{A}$ in $\overline{\mathcal{A}/\mathcal{G}}$ is completely characterized by the homomorphism $\hat{H}_{\bar{A}}$ from the hoop group $\mathcal{HG}$ to $SU(2)$, which is unique up to the freedom $\hat{H}_{\bar{A}} \mapsto g_0^{-1} \hat{H}_{\bar{A}} \, g_0$ for some $g_0 \in SU(2)$. Now suppose we are given n independent loops, $\beta_1, \ldots, \beta_n$. Denote by S^* the subgroup of the hoop group $\mathcal{HG}$ they generate. Using this S^*, let us introduce the following equivalence relation on $\overline{\mathcal{A}/\mathcal{G}}$: $\bar{A}_1 \sim \bar{A}_2$ iff $\hat{H}_{\bar{A}_1}(\tilde{\gamma}) = g_0^{-1} \hat{H}_{\bar{A}_2}(\tilde{\gamma}) \, g_0$ for all $\tilde{\gamma} \in S^*$ and some (hoop-independent) $g_0 \in SU(2)$. Denote by $\pi(S^*)$ the projection from $\overline{\mathcal{A}/\mathcal{G}}$ onto the quotient space $[\overline{\mathcal{A}/\mathcal{G}}]/\sim$. The idea is to consider functions f on $\overline{\mathcal{A}/\mathcal{G}}$ which are pullbacks under $\pi(\mathcal{S}^\star)$ of functions $\hat{f}$ on $[\overline{\mathcal{A}/\mathcal{G}}]/\sim$ and define the integral of f on $\overline{\mathcal{A}/\mathcal{G}}$ to be equal to the integral of $\hat{f}$ on $[\overline{\mathcal{A}/\mathcal{G}}]/\sim$. However, for this strategy to work, $[\overline{\mathcal{A}/\mathcal{G}}]/\sim$ should have a simple structure that one can control. Fortunately, this is the case:

Lemma 4.1 *Let $\mathcal{S}^\star$ be a subgroup of the hoop group which is generated by a finite number of independent hoops. Then,*
(a) There is a natural bijection

$$[\overline{\mathcal{A}/\mathcal{G}}]/\sim \;\; \to \;\; \mathrm{Hom}(\mathcal{S}^\star, SU(2))/\mathrm{Ad}$$

(b) For each choice of independent generators $\{\beta_1, \ldots, \beta_n\}$ of $\mathcal{S}^\star$, there is a bijection

$$[\overline{\mathcal{A}/\mathcal{G}}]/\sim \;\; \to \;\; [SU(2)]^n/\mathrm{Ad}$$

(c) Given $\mathcal{S}^\star$ as in (a), the topology induced on $[\overline{\mathcal{A}/\mathcal{G}}]/\sim$ from $[SU(2)]^n/\mathrm{Ad}$ according to (b) is independent of the choice of free generators of $\mathcal{S}^\star$.

Proof By definition of the equivalence relation $\sim$, every $\{\bar{A}\}$ in $[\overline{\mathcal{A}/\mathcal{G}}]/\sim$ is in 1–1 correspondence with the restriction to $\mathcal{S}^\star$ of an equivalence class $\{\hat{H}_{\bar{A}}\}$ of homomorphisms from the full hoop group $\mathcal{HG}$ to $SU(2)$. Now, it follows from Lemma 3.3 that every homomorphism from $\mathcal{S}^\star$ to $SU(2)$ extends to a homomorphism from the full hoop group to $SU(2)$. Therefore, there is a 1–1 correspondence between equivalence classes $\{A\}$ and equivalence classes of homomorphisms from $\mathcal{S}^\star$ to $SU(2)$ (where, as usual, two are equivalent if they differ by the adjoint action of $SU(2)$).

The statement (b) follows automatically from (a): the map in (b) is given by the values taken by an element of $\mathrm{Hom}(\mathcal{S}^\star, SU(2))$ on the chosen generators. The proof of (c) is a simple exercise. $\quad\square$

Some remarks about this result are in order:

1. Although we began with independent loops $\{\beta_1, \ldots, \beta_n\}$, the equivalence relation we introduced makes direct reference only to the subgroup $\mathcal{S}^\star$ of the hoop group they generate. This is similar to the situation in the linear case where one may begin with a set of linearly independent elements $s_1^*, \ldots, s_n^*$ of V^*, let them generate a subspace $S^\star$ and then introduce the equivalence relation on S using $S^\star$ directly.

2. However, a specific basis is often used in subsequent constructions. In particular, cylindrical measures are generally introduced as follows. One fixes a basis in $S^\star$, uses it to define an isomorphism between $S = V/\sim$ and $\mathbf{R}^n$, introduces a measure on $\mathbf{R}^n$ for each n and uses the isomorphism to define measures $d\tilde{\mu}$ on spaces S. These, in turn, define a cylindrical measure $d\mu$ on V through (4.1). However, since $S^\star$ admits other bases, at the end one must make sure that

the final measure $d\mu$ on V is well-defined, i.e., is insensitive to the specific choice of basis made in its definition. In the case now under consideration, we will follow a similar strategy. As we just showed in Lemma 4.1, given a set of independent loops which generate $\mathcal{S}^\star$, we can identify $\overline{[\mathcal{A}/\mathcal{G}]}/\sim$ with $\overline{[SU(2)]^n}/\text{Ad}$. Therefore, we can define a cylindrical measure on $\overline{\mathcal{A}/\mathcal{G}}$ by first choosing suitable measures on $[SU(2)]^n/\text{Ad}$ for each n, using the isomorphism to define measures on $[\mathcal{A}/\mathcal{G}]/\sim$ and showing finally that the final integrals are insensitive to the initial choices of generators of $\mathcal{S}^\star$.

3. The equivalence relation $\sim$ of Lemma 4.1 says that two generalized connections are equivalent if their actions on the group $\mathcal{S}^\star$, generated by a finite number of independent loops, are indistinguishable. It follows from Lemma 3.3 that each equivalence class $\{\bar{A}\}$ contains a regular connection A. Finally, if A and A' are two regular connections whose pullbacks to all β_i are the same for $i = 1, \ldots, n$, then they are equivalent (where β_i is any loop in the hoop class $\tilde{\beta}_i$). Thus, in effect, the projection $\pi(\mathcal{S}^\star)$ maps the domain space $\overline{\mathcal{A}/\mathcal{G}}$ of the continuum quantum theory to the domain space of a lattice gauge theory where the lattice is formed by the n loops $\beta_1, \ldots, \beta_n$. The initial choice of the independent hoops is, however, arbitrary and it is this arbitrariness that is responsible for the richness of the continuum theory.

We can now define cylindrical functions on $\overline{\mathcal{A}/\mathcal{G}}$. Let $\mathcal{S}^\star$ be a subgroup of the hoop group which is generated by a finite number of independent loops. Let us set

$$\mathcal{B}_{\mathcal{S}^\star} := \text{Hom}(\mathcal{S}^\star, SU(2))/\text{Ad}$$

Using Lemma 4.1 we shall regard $\pi(\mathcal{S}^\star)$ as a projection map from $\overline{\mathcal{A}/\mathcal{G}}$ onto $\mathcal{B}_{\mathcal{S}^\star}$ and parametrize $\mathcal{B}_{\mathcal{S}^\star}$ by $[SU(2)]^n/\text{Ad}$. For the topology on $\mathcal{B}_{\mathcal{S}^\star}$ we will use the one defined by its identification with $[SU(2)]^n/\text{Ad}$. By a *cylindrical function* f on $\overline{\mathcal{A}/\mathcal{G}}$, we shall mean the pullback to $\overline{\mathcal{A}/\mathcal{G}}$ under $\pi(\mathcal{S}^\star)$ of a continuous function $\tilde{f}$ on $\mathcal{B}_{\mathcal{S}^\star}$, for *some* subgroup $\mathcal{S}^\star$ of the hoop group which is generated by a finite number of independent loops. These are the functions we want to integrate. Let us note an elementary property of cylindrical functions which will be used repeatedly. Let $\mathcal{S}^\star$ and $\mathcal{S}^{\star\prime}$ be two finitely generated subgroups of the hoop group and such that $\mathcal{S}^{\star\prime} \subseteq \mathcal{S}^\star$. Then, it is easy to check that if a function f on $\overline{\mathcal{A}/\mathcal{G}}$ is cylindrical with respect to $\mathcal{S}^{\star\prime}$, it is also cylindrical with respect to $\mathcal{S}^\star$. Furthermore, there is now a natural projection from $\mathcal{B}_{\mathcal{S}^\star}$ onto $\mathcal{B}_{\mathcal{S}^{\star\prime}}$ (given by the projection $\text{Hom}(\mathcal{S}^\star, SU(2)) \to \text{Hom}(\mathcal{S}^{\star\prime}, SU(2)))$ and $\tilde{f}$ is the pullback of $\tilde{f}'$ via this projection.

In the remainder of this subsection, we will analyse the structure of the space $\mathcal{C}$ of cylindrical functions. First we have:

Lemma 4.2 *$\mathcal{C}$ has the structure of a normed $\star$–algebra with respect to the operations of taking linear combinations, multiplication, complex conjugation and sup norm of cylindrical functions.*

Proof It is clear by definition of $\mathcal{C}$ that if $f \in \mathcal{C}$, then any complex multiple λf as well as the complex conjugate $\bar{f}$ also belongs to $\mathcal{C}$. Next, given two elements, f_i with $i = 1, 2$, of $\mathcal{C}$, we will show that their sum and their product also belong to $\mathcal{C}$. Let $\mathcal{S}_i^\star$ be finitely generated subgroups of the hoop group $\mathcal{HG}$ such that f_i are the pullbacks to $\overline{\mathcal{A}/\mathcal{G}}$ of continuous functions $\tilde{f}_i$ on $\mathcal{B}_{\mathcal{S}_i^\star}$. Consider the sets of n_1 and n_2 independent hoops generating respectively $\mathcal{S}_1^\star$ and $\mathcal{S}_2^\star$. While the first n_1 and the last n_2 hoops are independent, there may be relations between hoops belonging to the first set and those belonging to the second. Using the technique of Section 3.1, find the set of independent hoops, say, $\tilde{\beta}_1, \ldots, \tilde{\beta}_n$ which generate the given $n_1 + n_2$ hoops and denote by $\mathcal{S}^\star$ the subgroup of the hoop group generated by $\tilde{\beta}_1, \ldots, \tilde{\beta}_n$. Then, since $\mathcal{S}_i^\star \subseteq \mathcal{S}^\star$, it follows that f_i are cylindrical also with respect to $\mathcal{S}^\star$. Therefore, $f_1 + f_2$ and $f_1 \cdot f_2$ are also cylindrical with respect to $\mathcal{S}^\star$. Thus $\mathcal{C}$ has the structure of a $\star$-algebra. Finally, since $\mathcal{B}_{\mathcal{S}^\star}$ (being homeomorphic to $[SU(2)]^n/\mathrm{Ad}$) is compact for any $\mathcal{S}^\star$, and since elements of $\mathcal{C}$ are pullbacks to $\overline{\mathcal{A}/\mathcal{G}}$ of continuous functions on $\mathcal{B}_{\mathcal{S}^\star}$, it follows that the cylindrical functions are bounded. We can therefore use the sup norm to endow $\mathcal{C}$ with the structure of a normed $\star$-algebra. $\square$

We will use the sup norm to complete $\mathcal{C}$ and obtain a C*-algebra $\overline{\mathcal{C}}$. A natural class of functions one would like to integrate on $\overline{\mathcal{A}/\mathcal{G}}$ is given by the (Gel'fand transforms of the) traces of holonomies. The question then is if these functions are contained in $\overline{\mathcal{C}}$. Not only is the answer affirmative, but in fact the traces of holonomies generate all of $\overline{\mathcal{C}}$. More precisely, we have:

Theorem 4.3 *The C*-algebra $\overline{\mathcal{C}}$ is isomorphic to the C*-algebra $\overline{\mathcal{HA}}$.*

Proof Let us begin by showing that $\overline{\mathcal{HA}}$ is embedded in $\overline{\mathcal{C}}$. Given an element F with $F(A) = \sum a_i T_{\tilde{\alpha}_i}(A)$ of $\overline{\mathcal{HA}}$, its Gel'fand transform is a function f on the spectrum $\overline{\mathcal{A}/\mathcal{G}}$, given by $f(\bar{A}) = \sum a_i \bar{A}(\tilde{\alpha}_i)$. Consider the set of hoops $\tilde{\alpha}_1, \ldots, \tilde{\alpha}_k$ that feature in the sum, decompose them into independent hoops $\tilde{\beta}_1, \ldots, \tilde{\beta}_n$ as in Section 3.1, and denote by $\mathcal{S}^\star$ the subgroup of the hoop group they generate. Then, it is clear that the function f is the pullback via $\pi(\mathcal{S}^\star)$ of a continuous function $\tilde{f}$ on $\mathcal{B}_{\mathcal{S}^\star}$. Hence we have $f \in \mathcal{C}$. Since the norms of F in $\overline{\mathcal{HA}}$ and f in $\overline{\mathcal{C}}$ are given by

$$\|F\| = \sup_{A \in \mathcal{A}} |F(A)| \quad \text{and} \quad \|f\| = \sup_{\bar{A} \in \overline{\mathcal{A}/\mathcal{G}}} |f(\bar{A})|,$$

since the restriction of f to $\mathcal{A}/\mathcal{G}$ coincides with F, and since $\mathcal{A}/\mathcal{G}$ is embedded densely in $\overline{\mathcal{A}/\mathcal{G}}$, it is clear that the map from F to f is isometric. Finally, since $\overline{\mathcal{HA}}$ is obtained by a C*-completion of the space of functions of the form F, we have the result that $\overline{\mathcal{HA}}$ is embedded in the C*-algebra $\overline{\mathcal{C}}$.

Next, we will show that there is also an inclusion in the opposite direction. We first note a fact about $SU(2)$. Fix n elements $(g_1, \ldots, g_n)$ of $SU(2)$. Then, from the knowledge of traces (in the fundamental representation) of all elements which belong to the group generated by these $g_i, i = 1, \ldots, n$, we can reconstruct $(g_1, \ldots, g_n)$ modulo an adjoint map. It therefore follows that the space of functions $\tilde{f}$ on $\mathcal{B}_{S^*}$ (considered as a copy of $[SU(2)]^n/\mathrm{Ad}$) which come from the functions F on $\overline{\mathcal{HA}}$ (using all the hoops $\tilde{\alpha}_i \in \mathcal{S}^*$) suffice to separate points of $\mathcal{B}_{S^*}$. Since this space is compact, it follows (from the Weierstrass theorem) that the C*-algebra generated by these projected functions is the entire C*-algebra of continuous functions on $\mathcal{B}_{S^*}$. Now, suppose we are given a cylindrical function f' on $\overline{\mathcal{A}/\mathcal{G}}$. Its projection under the appropriate $\pi(\mathcal{S}^*)$ is, by the preceeding result, contained in the projection of some element of $\overline{\mathcal{HA}}$. Thus, $\mathcal{C}$ is contained in $\overline{\mathcal{HA}}$.

Combining the two results, we have the desired result: the C*-algebras $\overline{\mathcal{HA}}$ and $\overline{\mathcal{C}}$ are isomorphic.   □

Remark Theorem 4.3 suggests that from the very beginning we could have introduced the C*-algebra $\overline{\mathcal{C}}$ in place of $\overline{\mathcal{HA}}$. This is indeed the case. More precisely, we could have proceeded as follows. Begin with the space $\mathcal{A}/\mathcal{G}$, and introduce the notion of hoops and hoop decomposition in terms of independent hoops. Then given a subgroup $\mathcal{S}^*$ of $\mathcal{HG}$ generated by a finite number of independent hoops, we could have introduced an equivalence relation on $\mathcal{A}/\mathcal{G}$ (*not* $\overline{\mathcal{A}/\mathcal{G}}$ which we do not yet have!) as follows: $A_1 \sim A_2$ iff $H(\tilde{\gamma}, A_1) = g^{-1}H(\tilde{\gamma}, A_2)g$ for all $\tilde{\gamma} \in \mathcal{S}^*$ and some (hoop independent) $g \in SU(2)$. It is then again true (due to Lemma 3.3) that the quotient $[\mathcal{A}/\mathcal{G}]/\sim$ is isomorphic to $\mathcal{B}_{S^*}$. (This is true inspite of the fact that we are using $\mathcal{A}/\mathcal{G}$ rather than $\overline{\mathcal{A}/\mathcal{G}}$ because, as remarked immediately after Lemma 4.1, each $\{\bar{A}\} \in [\overline{\mathcal{A}/\mathcal{G}}]/\sim$ contains a regular connection.) We can therefore define cylindrical functions, but now on $\mathcal{A}/\mathcal{G}$ (rather than $\overline{\mathcal{A}/\mathcal{G}}$) as the pullbacks of continuous functions on $\mathcal{B}_{S^*}$. These functions have a natural C*-algebra structure. We can use it as the starting point in place of the A–I holonomy algebra. While this strategy seems natural at first from an analysis viewpoint, it has two drawbacks, both stemming from the fact that the Wilson loops are now assigned a secondary role. For physical applications, this is unsatisfactory since Wilson loops are the basic observables of gauge theories. From a mathematical viewpoint, the relation between knot/link invariants and measures on the spectrum of the algebra would now be obscure. Nonetheless, it is good to keep in mind that this alternate strategy *is* available as it may make other issues more transparent.

4.2   A natural measure

In this sub–section, we will discuss the issue of integration of cylindrical functions on $\overline{A/G}$. Our main objective is to introduce a natural, faithful, diffeomorphism invariant measure on $\overline{A/G}$.

As mentioned in the remarks following Lemma 4.1, the idea is similar to the one used in the case of topological vector spaces. Thus, for each subgroup $S^\star$ of the hoop group which is generated by a finite number of independent hoops, we will equip the space $\mathcal{B}_{S^\star}$ with a measure $d\mu_{S^\star}$ and, as in (4.1), set

$$\int_{\overline{A/G}} d\mu\, f := \int_{\mathcal{B}_{S^\star}} d\mu_{S^\star}\, \tilde{f} \tag{4.3}$$

where f is any cylindrical function compatible with $S^\star$. It is clear that for the integral to be well–defined, the set of measures we choose on different $\mathcal{B}_{S^\star}$ should be compatible so that the analog of (4.2) holds. We will now exhibit a natural choice for which the compatibility is automatically satisfied.

Denote by $d\mu$ the normalized Haar measure on $SU(2)$. It naturally induces a measure on $[SU(2)]^n$ which is invariant under the adjoint action of $SU(2)$ and therefore projects down unambiguously to a measure $d\mu(n)$ on $[SU(2)]^n/\text{Ad}$. Next, consider an arbitrary group $S^\star$ which is generated by n independent loops. Choose a set of independent generators $\{\beta_1, \ldots, \beta_n\}$ and use the corresponding map $[SU(2)]^n/\text{Ad} \to \mathcal{B}_{S^\star}$ to transport the measure $d\mu(n)$ to a measure $d\mu_{S^\star}$ on $\mathcal{B}_{S^\star}$. We will show that the measure $d\mu_{S^\star}$ is insensitive to the specific choice of the generators of $S^\star$ used in its construction and that the resulting family of measures on the various $\mathcal{B}_{S^\star}$ satisfies the appropriate compatibility conditions. We will then explore the properties of the resulting *cylindrical measure* on $\overline{A/G}$. Our construction is natural since $SU(2)$ comes equipped with the Haar measure. In particular, we have not introduced any additional structure on the underlying 3-manifold Σ (on which the initial connections A are defined); hence the resulting cylindrical measure will be automatically invariant under the action of $\text{Diff}(\Sigma)$.

Theorem 4.4   *a) Let f be a cylindrical function on $\overline{A/G}$ with respect to two subgroups $S_i^\star$, $i = 1, 2$, of the hoop group, each of which is generated by a finite number of independent hoops. Fix a set of generators in each subgroup and denote by $d\mu_{S_i^\star}$ the corresponding measures on $\mathcal{B}_{S_i^\star}$. Then, if $\tilde{f}_i$ denote the projections of f to $\mathcal{B}_{S_i^\star}$, we have:*

$$\int_{\mathcal{B}_{S_1^\star}} d\mu_{S_1^\star}\, \tilde{f}_1 = \int_{\mathcal{B}_{S_2^\star}} d\mu_{S_2^\star}\, \tilde{f}_2. \tag{4.4}$$

In particular, if $S_1^\star = S_2^\star$, the integral is independent of the choice of generators used in its definition.

b) The functional $v: \mathcal{HA} \to \mathbf{C}$ *defined below extends to a linear, strictly positive and Diff(Σ)-invariant functional on* $\overline{\mathcal{HA}}$:

$$v(F) = \int_{\overline{\mathcal{A}/\mathcal{G}}} d\mu \; f \qquad (4.5)$$

where $d\mu$ *is defined by (4.3) and where the cylindrical function* f *on* $\overline{\mathcal{A}/\mathcal{G}}$ *is the Gel'fand transform of the element* F *of* $\overline{\mathcal{HA}}$; *and,*
c) The cylindrical 'measure' $d\mu$ *defined by (4.3) is a genuine, regular and strictly positive measure on* $\overline{\mathcal{A}/\mathcal{G}}$. *Furthermore,* $d\bar{\mu}$ *is* Diff(Σ) *invariant.*

Proof Fix n_i independent hoops that generate $\mathcal{S}_i^*$ and, as in the proof of Theorem 4.3, use the construction of Section 3.1 to carry out the decomposition of these $n_1 + n_2$ hoops in terms of independent hoops, say $\tilde{\beta}_1, \ldots, \tilde{\beta}_n$. Thus, the original $n_1 + n_2$ hoops are contained in the group $\mathcal{S}^*$ generated by the $\tilde{\beta}_i$. Clearly, $\mathcal{S}_i^* \subseteq \mathcal{S}^*$. Hence, the given f is cylindrical also with respect to $\mathcal{S}^*$, i.e. is the pullback of a function $\tilde{f}$ on $\mathcal{B}_{\mathcal{S}^*}$. We will now establish (4.4) by showing that each of the two integrals in that equation equals the analogous integral of $\tilde{f}$ over $\mathcal{B}_{\mathcal{S}^*}$.

Note first that, since the generators have been fixed, we can ignore the distinction between $\mathcal{B}_{\mathcal{S}_i^*}$ and $[SU(2)]^{n_i}/\mathrm{Ad}$ and between $\mathcal{B}_{\mathcal{S}^*}$ and $[SU(2)]^n/\mathrm{Ad}$. It will also be convenient to regard functions on $[SU(2)]^n/\mathrm{Ad}$ as functions on $[SU(2)]^n$ which are invariant under the adjoint $SU(2)$ action. Thus, instead of carrying out integrals over $\mathcal{B}_{\mathcal{S}_i^*}$ and $\mathcal{B}_{\mathcal{S}^*}$ of functions on these spaces, we can integrate the lifts of these functions to $[SU(2)]^{n_i}$ and $[SU(2)]^n$ respectively.

To deal with quantities with suffixes 1 and 2 simultaneously, for notational simplicity we will let a prime stand for *either* the suffix 1 *or* 2. Then, in particular, we have finitely generated subgroups $\mathcal{S}^*$ and $(\mathcal{S}^*)'$ of the hoop group, with $(\mathcal{S}^*)' \subset \mathcal{S}^*$, and a function f which is cylindrical with respect to both of them. The generators of $\mathcal{S}^*$ are $\tilde{\beta}_1, \ldots, \tilde{\beta}_n$ while those of $(\mathcal{S}^*)'$ are $\tilde{\beta}_1', \ldots, \tilde{\beta}_{n'}'$ (with $n' < n$). From the construction of $\mathcal{S}^*$ (Section 3.1) it follows that every hoop in the second set can be expressed as a composition of the hoops in the first set and their inverses. Furthermore, since the hoops in each set are independent, the construction implies that for each $i' \in \{1, \ldots, n'\}$, there exists $K(i') \in \{1, \ldots, n\}$ with the following properties: (i) if $i' \neq j'$, $K(i') \neq K(j')$; and, (ii) in the decomposition of $\tilde{\beta}_{i'}'$ in terms of unprimed hoops, the hoop $\tilde{\beta}_{K(i')}$ (or its inverse) appears exactly once and if $i' \neq j'$, the hoop $\tilde{\beta}_{K(j')}$ does not appear at all. For simplicity, let us assume that the orientation of the primed hoops is such that it is $\beta_{K(i')}$ rather than its inverse that features in the decomposition of $\beta_{i'}'$. Next, let us denote by $(g_1, \ldots, g_n)$ a point of the quotient $[SU(2)]^n$ which is projected to $(g_1', \ldots, g_{n'}')$ in $[SU(2)]^{n'}$. Then $g_{i'}'$ is expressed in terms of g_i as $g_{i'}' = ..g_{K(i')}..$, where .. denotes a product of elements of the type g_m where $m \neq K(j)$ for any j. Since the given function f on $\overline{\mathcal{A}/\mathcal{G}}$ is a pullback of a function $\tilde{f}$ on $[SU(2)]^n$ as well as of a function $\tilde{f}'$ on $[SU(2)]^{n'}$, it follows that

$$\tilde{f}(g_1, \ldots, g_n) = \tilde{f}'(g_1', \ldots, g_{n'}') = \tilde{f}'(\cdot\cdot g_{K(1)}\cdot\cdot, \ldots, \cdot\cdot g_{K(n')}\cdot\cdot).$$

Hence,

$$\int d\tilde{\mu}_1 \cdots \int d\tilde{\mu}_n \, \tilde{f}(g_1, \ldots, g_n) =$$

$$\int \prod_{m \neq K(1)\ldots K(n')} d\tilde{\mu}_m \left(\int d\tilde{\mu}_{K(1)} \cdots \int d\tilde{\mu}_{K(n')} \tilde{f}'(\cdot\cdot g_{K(1)}\cdot\cdot, \ldots, \cdot\cdot g_{K(n')}\cdot\cdot) \right)$$

$$= \int \prod_{m=n'+1}^{n} d\tilde{\mu}_m \left(\int d\tilde{\mu}_1 \cdots \int d\tilde{\mu}_{n'} \tilde{f}'(g_1, \ldots, g_{n'}) \right)$$

$$= \int d\tilde{\mu}_1 \cdots \int d\tilde{\mu}_{n'} \tilde{f}'(g_1, \ldots, g_{n'}),$$

where, in the first step, we simply rearranged the order of integration, in the second step we used the invariance property of the Haar measure and simplified notation for dummy variables being integrated, and, in the third step, we used the fact that, since the measure is normalized, the integral of a constant produces just that constant[8]. This establishes the required result (4.4).

Next, consider the 'vacuum expectation value functional' v on $\mathcal{HA}$ defined by (4.5). The continuity and linearity of v follow directly from its definition. That it is positive, i.e. satisfies $v(F^\star F) \geq 0$, is equally obvious. In fact, a stronger result holds: if $F \neq 0$, then $v(F^\star F) > 0$ since the value $v(F^\star F)$ of v on $F^\star F$ is obtained by integrating a non-negative function $(\tilde{f})^\star \tilde{f}$ on $[SU(2)]^n/\mathrm{Ad}$ with respect to a regular measure. Thus, v is *strictly* positive. Since $\mathcal{HA}$ is dense in $\overline{\mathcal{HA}}$ it follows that the functional extends to $\overline{\mathcal{HA}}$ by continuity.

To see that we have a regular measure on $\overline{\mathcal{A}/\mathcal{G}}$, recall first that the C*-algebra of cylindrical functions is naturally isomorphic with the Gel'fand transform of the C*-algebra $\overline{\mathcal{HA}}$, which, in turn is the C*-algebra of all continuous functions on the compact Hausdorff space $\overline{\mathcal{A}/\mathcal{G}}$. The 'vacuum expectation value function' v is therefore a continuous linear functional on the algebra of all continuous functions on $\overline{\mathcal{A}/\mathcal{G}}$ equipped with the L^∞ (i.e. sup) norm. Hence, by the Riesz–Markov theorem, it follows that

---

[8]In the simplest case, with $n = 2$ and $n' = 1$, we would for example have $\tilde{f}(g_1, g_2) = \tilde{f}'(g_1 \cdot g_2) \equiv \tilde{f}'(g')$. Then $\int d\mu(g_1) \int d\mu(g_2)\tilde{f}(g_1, g_2) = \int d\mu(g_1) \int d\mu(g_2)\tilde{f}'(g_1 \cdot g_2) = \int d\mu(g')\tilde{f}(g')$, where in the second step we have used the invariance and normalization properties of the Haar measure.

$v(f) \equiv \int d\mu f$ is in fact the integral of the continuous function f on $\overline{\mathcal{A}/\mathcal{G}}$ with respect to a regular measure. This is just our cylindrical measure.

Finally, since our construction did not require the introduction of any extra structure on Σ (such as a metric or a volume element), it follows that the functional v and the measure $d\mu$ are Diff(Σ) invariant. $\qquad\square$

We will refer to $d\mu$ as the 'induced Haar measure' on $\overline{\mathcal{A}/\mathcal{G}}$. We now explore some properties of this measure and of the resulting representation of the holonomy C*-algebra $\overline{\mathcal{HA}}$.

First, the representation of $\overline{\mathcal{HA}}$ can be constructed as follows. The Hilbert space is $L^2(\overline{\mathcal{A}/\mathcal{G}}, d\mu)$ and the elements F of $\overline{\mathcal{HA}}$ act by multiplication so that we have $F\Psi = f \cdot \Psi$ for all $\Psi \in L^2(\overline{\mathcal{A}/\mathcal{G}}, d\mu)$, where $f \in \mathcal{C}$ is the Gel'fand transform of F. This representation is cyclic and the function Ψ_0 defined by $\Psi_0(\bar{A}) = 1$ on $\overline{\mathcal{A}/\mathcal{G}}$ is the 'vacuum' state. Every generator T_α of the holonomy C*-algebra is represented by a bounded, self-adjoint operator on $L^2(\overline{\mathcal{A}/\mathcal{G}}, d\mu)$.

Second, it follows from the strictness of the positivity of the measure that the resulting representation of the C*-algebra $\overline{\mathcal{HA}}$ is *faithful*. To our knowledge, this is the first faithful representation of the holonomy C*-algebra that has been constructed explicitly. Finally, the diffeomorphism group Diff(Σ) of Σ has an induced action on the C*-algebra $\overline{\mathcal{HA}}$ and hence also on its spectrum $\overline{\mathcal{A}/\mathcal{G}}$. Since the measure $d\mu$ on $\overline{\mathcal{A}/\mathcal{G}}$ is invariant under this action, the Hilbert space $L^2(\overline{\mathcal{A}/\mathcal{G}}, d\mu)$ carries a unitary representation of Diff(Σ). Under this action, the 'vacuum state' Ψ_0 is left invariant.

Third, restricting the values of v to the generators $T_{\tilde{\alpha}} \equiv [\alpha]_K$ of $\overline{\mathcal{HA}}$, we obtain the generating functional $\Gamma(\alpha)$ of this representation: $\Gamma(\alpha) := \int d\mu \bar{A}(\tilde{\alpha})$. This functional on the loop space $\mathcal{L}_{x^0}$ is diffeomorphism invariant since the measure $d\mu$ is. It thus defined a generalized knot invariant. We will see below that, unlike the standard invariants, $\Gamma(\alpha)$ vanishes on all smoothly embedded loops. Its action is non-trivial only when the loop is retraced, i.e. only on generalized loops.

Let us conclude this section by indicating how one computes the integral in practice. Fix a loop $\alpha \in \mathcal{L}_{x^0}$ and let

$$[\alpha] = [\beta_{i_1}]^{\epsilon_1} \cdots [\beta_{i_N}]^{\epsilon_N} \qquad (4.6)$$

be the minimal decomposition of α into independent loops $\beta_1, \ldots, \beta_n$ as in Section 3.1, where $\epsilon_k \in \{-1, 1\}$ keeps track of the orientation and $i_k \in \{1, \ldots, n\}$. (We have used the double suffix i_k because any one independent loop $\beta_1, \ldots, \beta_n$ can arise more than once in the decomposition. Thus, $N \geq n$.) The loop α defines an element $T_{\tilde{\alpha}}$ of the holonomy C*-algebra $\overline{\mathcal{HA}}$ whose Gel'fand transform $t_{\tilde{\alpha}}$ is a function on $\overline{\mathcal{A}/\mathcal{G}}$. This is the cylindrical function we wish to integrate. Now, the projection map of Lemma 4.1 assigns to $t_{\tilde{\alpha}}$ the function $\tilde{t}_{\tilde{\alpha}}$ on $[SU(2)]^n/\mathrm{Ad}$ given by

$$\tilde{t}_{\tilde{\alpha}}(g_1, \ldots, g_n) := \mathrm{Tr}\ (g_{i_1}^{\epsilon_1} \cdots g_{i_N}^{\epsilon_N}).$$

Thus, to integrate $t_{\tilde{\alpha}}$ on $\overline{\mathcal{A}/\mathcal{G}}$, we need to integrate this $\tilde{t}_{\tilde{\alpha}}$ on $[SU(2)]^n/\mathrm{Ad}$. Now, there exist in the literature useful formulae for integrating polynomials of the 'matrix element functions' over $SU(2)$ (see, e.g., [14]). These can be now used to evaluate the required integrals on $[SU(2)]^n/\mathrm{Ad}$. In particular, one can readily show the following result:

$$\int_{\overline{\mathcal{A}/\mathcal{G}}} d\mu \; t_{\tilde{\alpha}} = 0,$$

unless *every* loop β_i is repeated an even number of times in the decomposition (4.6) of α.

5   Discussion

In this paper, we completed the A–I program in several directions. First, by exploiting the fact that the loops are piecewise analytic, we were able to obtain a complete characterization of the Gel'fand spectrum $\overline{\mathcal{A}/\mathcal{G}}$ of the holonomy C*-algebra $\overline{\mathcal{HA}}$. A–I had shown that every element of $\overline{\mathcal{A}/\mathcal{G}}$ defines a homomorphism from the hoop group $\mathcal{HG}$ to $SU(2)$. We found that every homomorphism from $\mathcal{HG}$ to $SU(2)$ defines an element of the spectrum and two homomorphisms $\hat{H}_1$ and $\hat{H}_2$ define the same element of $\overline{\mathcal{A}/\mathcal{G}}$ only if they differ by an $SU(2)$ automorphism, i.e. if and only if $\hat{H}_2 = g^{-1} \cdot \hat{H}_1 \cdot g$ for some $g \in SU(2)$. Since this characterization is rather simple and purely algebraic, it is useful in practice. The second main result also intertwines the structure of the hoop group with that of the Gel'fand spectrum. Given a subgroup $\mathcal{S}^*$ of $\mathcal{HG}$ generated by a finite number, say n, of independent hoops, we were able to define a projection $\pi(\mathcal{S}^*)$ from $\overline{\mathcal{A}/\mathcal{G}}$ to the compact space $[SU(2)]^n/\mathrm{Ad}$. This family of projections enables us to define cylindrical functions on $\overline{\mathcal{A}/\mathcal{G}}$: these are the pullbacks to $\overline{\mathcal{A}/\mathcal{G}}$ of the continuous functions on $[SU(2)]^n/\mathrm{Ad}$. We analysed the space $\mathcal{C}$ of these functions and found that its completion $\overline{\mathcal{C}}$ has the structure of a C*-algebra which, moreover, is naturally isomorphic with the C*-algebra of all continuous functions on the compact Hausdorff space $\overline{\mathcal{A}/\mathcal{G}}$ and hence, via the inverse Gel'fand transform, also to the holonomy C*-algebra $\overline{\mathcal{HA}}$ with which we began. We then carried out a non-linear extension of the theory of cylindrical measures to integrate these cylindrical functions on $\overline{\mathcal{A}/\mathcal{G}}$. Since $\overline{\mathcal{C}}$ *is* the C*-algebra of all continuous functions on $\overline{\mathcal{A}/\mathcal{G}}$, the cylindrical measures in fact turn out to be regular measures on $\overline{\mathcal{A}/\mathcal{G}}$. Finally, we were able to introduce explicitly such a measure on $\overline{\mathcal{A}/\mathcal{G}}$ which is natural in the sense that it arises from the Haar measure on $SU(2)$. The resulting representation of the holonomy C*-algebra has several interesting properties. In particular, the representation is faithful and diffeomorphism invariant. Hence, its generating function Γ—which sends loops α (based at x^0) to real numbers $\Gamma(\alpha)$—defines a generalized knot invariant (generalized because we have allowed loops to have kinks, self-intersections, and self-overlaps). These constructions show that the A–I program can be realized in detail.

Having the induced Haar measure $d\mu$ on $\overline{\mathcal{A}/\mathcal{G}}$ at one's disposal, we can now make the ideas on the loop transform $\mathcal{T}$ of [3, 4] rigorous. The transform $\mathcal{T}$ sends states in the connection representation to states in the so-called loop representation. In terms of machinery introduced in this paper, a state Ψ in the connection representation is a square-integrable function on $\overline{\mathcal{A}/\mathcal{G}}$; thus $\Psi \in L^2(\overline{\mathcal{A}/\mathcal{G}}, d\mu)$. The transform $\mathcal{T}$ sends it to a function ψ on the space $\mathcal{HG}$ of hoops. (Sometimes, it is convenient to lift ψ canonically to the space $\mathcal{L}_{x^0}$ and regard it as a function on the loop space. This is the origin of the term 'loop representation'.) We have $\mathcal{T} \circ \Psi = \psi$ with

$$\psi(\tilde{\alpha}) = \int_{\overline{\mathcal{A}/\mathcal{G}}} d\mu \, \overline{t_{\tilde{\alpha}}(\overline{A})} \Psi(\overline{A}) = \langle t_{\tilde{\alpha}}, \Psi \rangle, \tag{5.1}$$

where $t_{\tilde{\alpha}}$ (with $t_{\tilde{\alpha}}(\overline{A}) = \overline{A}(\tilde{\alpha})$) is the Gel'fand transform of the trace of the holonomy function, $T_{\tilde{\alpha}}$, on $\mathcal{A}/\mathcal{G}$, associated with the hoop $\tilde{\alpha}$, and $\langle \cdot, \cdot \rangle$ denotes the inner product on $L^2(\overline{\mathcal{A}/\mathcal{G}}, d\mu)$. Thus, in the transform the traces of holonomies play the role of the 'integral kernel'. Elements of the holonomy algebra have a natural action on the connection states Ψ. Using $\mathcal{T}$, one can transform this action to the hoop states ψ. It turns out that the action is surprisingly simple and can be represented directly in terms of elementary operations on the hoop space. Consider a generator $T_{\tilde{\gamma}}$ of $\mathcal{HA}$. In the connection representation, we have $(T_{\tilde{\gamma}} \circ \Psi)(\overline{A}) = T_{\tilde{\gamma}}(\overline{A}) \cdot \psi(\overline{A})$. On the hoop states, this action translates to

$$(T_{\tilde{\gamma}} \circ \psi)(\tilde{\alpha}) = \frac{1}{2} \Big(\psi(\tilde{\alpha} \cdot \tilde{\gamma}) + \psi(\tilde{\alpha} \cdot \tilde{\gamma}^{-1}) \Big) \tag{5.2}$$

where $\tilde{\alpha} \cdot \tilde{\gamma}$ is the composition of hoops $\tilde{\alpha}$ and $\tilde{\gamma}$ in the hoop group. Thus, one can forgo the connection representation and work directly in terms of hoop states. We will show elsewhere that the transform also interacts well with the 'momentum operators' which are associated with closed strips, i.e. that these operators also have simple action on the hoop states.

The hoop states are especially well suited to deal with diffeomorphism (i.e. Diff(Σ)) invariant theories. In such theories, physical states are required to be invariant under Diff(Σ). This condition is awkward to impose in the connection representation, and it is difficult to control the structure of the space of resulting states. In the loop representation, by contrast, the task is rather simple: physical states depend only on generalized knot and link classes of loops. Because of this simplification and because the action of the basic operators can be represented by simple operations on hoops, the loop representation has proved to be a powerful tool in quantum general relativity in 3 and 4 dimensions.

The use of this representation, however, raises several issues which are still open. Perhaps the most basic of these is that, without referring back to the connection representation, we do not yet have a useful characterization of the hoop states which are images of connection states, i.e. of elements of $L^2(\overline{\mathcal{A}/\mathcal{G}}, d\mu)$. Neither do we have an expression of the inner product between hoop states. It would be extremely interesting to develop integration theory also over the hoop group $\mathcal{HG}$ and express the inner product between hoop states directly in terms of such integrals, without any reference to the connection representation. This may indeed be possible using again the idea of cylindrical functions, but now on $\mathcal{HG}$, rather than on $\overline{\mathcal{A}/\mathcal{G}}$, and exploiting, as before, the duality between these spaces. If this is achieved and if the integrals over $\overline{\mathcal{A}/\mathcal{G}}$ and $\mathcal{HG}$ can be related, we would have a non-linear generalization of the Plancherel theorem which establishes that the Fourier transform is an isomorphism between two spaces of L^2-functions. The loop transform would then become an extremely powerful tool.

Appendix A: C^1 loops and $U(1)$ connections

In the main body of the paper, we restricted ourselves to piecewise analytic loops. This restriction was essential in Section 3.1 for our decomposition of a given set of a finite number of loops into *independent* loops which in turn is used in every major result contained in this paper. The restriction is not as severe as it might first seem since every smooth 3-manifold admits a unique analytic structure. Nonetheless, it is important to find out if our arguments can be replaced by more sophisticated ones so that the main results would continue to hold even if the holonomy C*-algebra were constructed from piecewise smooth loops. In this appendix, we consider $U(1)$ connections (rather than $SU(2)$) and show that, in this theory, our main results do continue to hold for piecewise C^1 loops. However, the new arguments make a crucial use of the Abelian character of $U(1)$ and do not by themselves go over to non-Abelian theories. Nonetheless, this Abelian example is an indication that analyticity may not be indispensible even in the non-Abelian case.

The appendix is divided into three parts. The first is devoted to certain topological considerations which arise because 3-manifolds admit non-trivial $U(1)$-bundles. The second proves the analog of the spectrum theorem of Section 3. The third introduces a diffeomorphism-invariant measure on the spectrum and discusses some of its properties. Throughout this appendix, by loops we shall mean continuous, piecewise C^1 loops. We will use the same notation as in the main text but now those symbols will refer to the $U(1)$ theory. Thus, for example, $\mathcal{HG}$ will denote the $U(1)$-hoop group and $\mathcal{HA}$, the $U(1)$ holonomy $\star$-algebra.

A.1 Topological considerations

Fix a smooth 3-manifold Σ. Denote by $\mathcal{A}$ the space of all smooth $U(1)$ connections which belong to an appropriate Sobolev space. Now, unlike $SU(2)$

bundles, $U(1)$ bundles over 3-manifolds need not be trivial. The question therefore arises as to whether we should allow arbitrary $U(1)$ connections or restrict ourselves only to the trivial bundle in the construction of the holonomy C*-algebra. Fortunately, it turns out that both choices lead to the same C*-algebra. This is the main result of this subsection.

Denote by $\mathcal{A}^0$ the subspace of $\mathcal{A}$ consisting of connections on the trivial $U(1)$ bundle over Σ. As is common in the physics literature, we will identify elements A^0 of $\mathcal{A}^0$ with real-valued 1-forms. Thus, the holonomies defined by any $A^0 \in \mathcal{A}^0$ will be denoted by $H(\alpha, A^0) = \exp(i \oint_\alpha A^0) \equiv \exp(i\theta)$, where θ takes values in $(0, 2\pi)$ and depends on both the connection A^0 and the loop α. We begin with the following result:

Lemma A.1 *Given a finite number of loops, $\alpha_1, \ldots, \alpha_n$, for every $A \in \mathcal{A}$ there exists $A^0 \in \mathcal{A}^0$ such that*

$$H(\alpha_i, A) = H(\alpha_i, A^0), \quad \forall i \in \{1, \ldots, n\}.$$

Proof Suppose the connection A is defined on the bundle P. We will show first that there exists a local section s of P which contains the given loops α_i. To construct the section we use a map $\Phi : \Sigma \to S_2$ which generates the bundle $P(\Sigma, U(1))$ (see Trautman [15]). The map Φ carries each α_i into a loop $\check{\alpha}_i$ in S_2. Since the loops $\check{\alpha}_i$ cannot form a dense set in S_2, we may remove from the 2-sphere an open ball B which does not intersect any of $\check{\alpha}_i$. Now, since $S_2 - B$ is contractible, there exists a smooth section σ of the Hopf bundle $U(1) \to SU(2) \to S_2$,

$$\sigma : (S_2 - B) \mapsto SU(2).$$

Hence, there exists a section s of $P(\Sigma, U(1))$ defined on a subset $V = \Phi^{-1}(S_2 - B) \subset \Sigma$ which contains all the loops α_i. Having obtained this section s, we can now look for the connection A^0. Let ω be a globally defined, real 1-form on Σ such that $\omega|V = s^*A$. Hence, the connection $A^0 := \omega$ defines on the given loops α_i the same holonomy elements as A; i.e.

$$H(\alpha_i, A^0) = H(\alpha_i, A) \tag{A.1}$$

for all i. □

This lemma has several useful implications. We mention two that will be used immediately in what follows.

1. Definition of hoops: A priori, there are two equivalence relations on the space $\mathcal{L}_{x^0}$ of based loops that one can introduce. One may say that two loops are equivalent if the holonomies of all connections in $\mathcal{A}$ around them are the same, or, one could ask that the holonomies

be the same only for connections in $\mathcal{A}^0$. Lemma A.1 implies that the two notions are in fact equivalent; there is only one hoop group $\mathcal{HG}$. As in the main text, we will denote by $\tilde{\alpha}$ the hoop to which a loop $\alpha \in \mathcal{L}_{x^0}$ belongs.

2. Sup norm: Consider functions f on $\mathcal{A}$ defined by finite linear combinations of holonomies on $\mathcal{A}$ around hoops in $\mathcal{HG}$. (Since the holonomies themselves are complex numbers, the trace operation is now redundant.) We can restrict these functions to $\mathcal{A}^0$. Lemma A.1 implies that

$$\sup_{A \in \mathcal{A}} |f(A)| = \sup_{A^0 \in \mathcal{A}^0} |f(A^0)|. \tag{A.2}$$

As a consequence of these implications, we can use *either $\mathcal{A}$ or $\mathcal{A}^0$* to construct the holonomy C*-algebra; in spite of topological non-trivialities, there is only one $\overline{\mathcal{HA}}$. To construct this algebra we proceed as follows. Let $\mathcal{HA}$ denote the complex vector space of functions f on $\mathcal{A}^0$ of the form

$$f(A^0) = \sum_{j=1}^{n} a_j H(\alpha, A^0) \equiv \sum_{j=1}^{n} a_j \exp\left(i \oint_\alpha A^0\right), \tag{A.3}$$

where a_j are complex numbers. Clearly, $\mathcal{HA}$ has the structure of a $\star$-algebra with the product law $H(\alpha, A^0) \circ H(\beta, A^0) = H(\alpha \cdot \beta, A^0)$ and the $\star$-relation given by $(H(\alpha, A^0))^\star = H(\alpha^{-1}, A^0)$. Equip it with the sup norm and take the completion. The result is the required C*-algebra $\overline{\mathcal{HA}}$.

We conclude by noting another consequence of Lemma A.1: the (Abelian) hoop group $\mathcal{HG}$ has no torsion. More precisely, we have the following result:

Lemma A.2 *Let $\tilde{\alpha} \in \mathcal{HG}$. Then if $(\tilde{\alpha})^n = e$, the identity in $\mathcal{HG}$, for $n \in \mathbf{Z}$, then $\tilde{\alpha} = e$.*

Proof Since $(\tilde{\alpha})^n = e$, we have, for every $A^0 \in \mathcal{A}^0$,

$$H(\alpha, A^0) \equiv H(\alpha^n, \tfrac{1}{n} A^0) = 1,$$

where α is any loop in the hoop $\tilde{\alpha}$. Hence, by Lemma A.1, it follows that $\tilde{\alpha} = e$. □

Although the proof is more complicated, the analogous result holds also in the $SU(2)$ case treated in the main text. However, in that case, the hoop group is non-Abelian. As we will see below, it is the Abelian character of the $U(1)$-hoop group that makes the result of this lemma useful.

A.2 The Gel'fand spectrum of $\overline{\mathcal{HA}}$

We now want to obtain a complete characterization of the spectrum of the $U(1)$ holonomy C*-algebra $\overline{\mathcal{HA}}$ along the lines of Section 3. The analog of Lemma 3.2 is easy to establish: every element $\bar{A}$ of the spectrum defines a homomorphism $\hat{H}_{\bar{A}}$ from the hoop group $\mathcal{HG}$ to $U(1)$ (now given simply by $\hat{H}_{\bar{A}}(\tilde{\alpha}) = \bar{A}(\tilde{\alpha})$). This is not surprising. Indeed, it is clear from the discussion in Section 3 that this lemma continues to hold for piecewise smooth loops even in the full $SU(2)$ theory. However, the situation is quite different for Lemma 3.3 because there we made a crucial use of the fact that the loops were (continuous and) piecewise *analytic*. Therefore, we must now modify that argument suitably. An appropriate replacement is contained in the following lemma.

Lemma A.3 *For every homomorphism $\hat{H}$ from the hoop group $\mathcal{HG}$ to $U(1)$, every finite set of hoops $\{\tilde{\alpha}_1, \ldots, \tilde{\alpha}_k\}$, and every $\epsilon > 0$, there exists a connection $A^0 \in \mathcal{A}^0$ such that*

$$|\hat{H}(\tilde{\alpha}_i) - H(\alpha_i, A^0)| < \epsilon, \quad \forall i \in \{1, \ldots, k\}. \tag{A.4}$$

Proof Consider the subgroup $\mathcal{HG}(\tilde{\alpha}_1, \ldots, \tilde{\alpha}_n) \subset \mathcal{HG}$ that is generated by $\tilde{\alpha}_1, \ldots, \tilde{\alpha}_k$. Since $\mathcal{HG}$ is Abelian and since it has no torsion, $\mathcal{HG}(\tilde{\alpha}_1, \ldots, \tilde{\alpha}_k)$ is finitely and freely generated by some elements, say $\tilde{\beta}_1, \ldots, \tilde{\beta}_n$. Hence, if (A.4) is satisfied by $\tilde{\beta}_i$ for a sufficiently small ϵ' then it will also be satisfied by the given $\tilde{\alpha}_j$ for the given ϵ. Consequently, without loss of generality, we can assume that the hoops α_i are (weakly) $U(1)$-independent, i.e. that they satisfy the following condition:

$$\text{if } (\tilde{\alpha}_1)^{k_1} \cdots (\tilde{\alpha}_n)^{k_n} = e, \quad \text{then } k_i = 0 \; \forall i.$$

Now, given such hoops $\tilde{\alpha}_i$, the homomorphism $\hat{H} : \mathcal{HG} \mapsto U(1)$ defines a point $(\hat{H}(\tilde{\alpha}_1), \ldots, \hat{H}(\tilde{\alpha}_k)) \in [U(1)]^k$. On the other hand, there also exists a map from $\mathcal{A}^0$ to $[U(1)]^k$, which can be expressed as a composition:

$$\mathcal{A}^0 \ni A^0 \mapsto \left(\oint_{\alpha_1} A^0, \ldots, \oint_{\alpha_k} A^0 \right) \mapsto \left(e^{i \oint_{\alpha_1} A^0}, \ldots, e^{i \oint_{\alpha_k} A^0} \right) \in [U(1)]^k.$$
$$\tag{A.5}$$

The first map in (A.5) is linear and its image, $V \subset R^k$, is a vector space. Let $0 \neq \mathbf{m} = (m_1, \ldots, m_k) \in \mathbf{Z}^k$. Denote by $V_{\mathbf{m}}$ the subspace of V orthogonal to $\mathbf{m}$. Now if $V_{\mathbf{m}}$ were to equal V then we would have

$$(\tilde{\alpha}_1)^{m_1} \cdots (\tilde{\alpha}_k)^{m_k} = 1,$$

which would contradict the assumption that the given set of hoops is independent. Hence we necessarily have $V_{\mathbf{m}} < V$. Furthermore, since a countable union of subsets of measure zero has measure zero, it follows that

$$\bigcup_{\mathbf{m} \in \mathbf{Z}^k} V_{\mathbf{m}} < V.$$

This strict inequality implies that there exists a connection $B^0 \in \mathcal{A}^0$ such that

$$\mathbf{v} := \left(\oint_{\alpha_1} B^0, \ldots, \oint_{\alpha_n} B^0 \right) \in \left(V - \bigcup_{\mathbf{m} \in \mathbf{Z}^k} V_{\mathbf{m}} \right).$$

In other words, B^0 is such that every ratio v_i/v_j of two different components of $\mathbf{v}$ is irrational. Thus, the line in $\mathcal{A}^0$ defined by B^0 via $A^0(t) = tB^0$ is carried by the map $(A.5)$ into a line which is dense in $[U(1)]^k$. This ensures that there exists an A^0 on this line which has the required property $(A.4)$.

□

Armed with this substitute of Lemma 3.3, it is straightforward to show the analog of Lemma 3.4. Combining these results, we have the $U(1)$ spectrum theorem:

Theorem A.4 *Every $\bar{A}$ in the spectrum of $\overline{\mathcal{HA}}$ defines a homomorphism $\hat{H}$ from the hoop group $\mathcal{HG}$ to $U(1)$, and, conversely, every homomorphism $\hat{H}$ defines an element $\bar{A}$ of the spectrum such that $\hat{H}(\tilde{\alpha}) = \bar{A}(\tilde{\alpha})$. This is a 1–1 correspondence.*

A.3 A natural measure

Results presented in the previous subsection imply that one can again introduce the notion of cylindrical functions and measures. In this subsection, we will exhibit a natural measure. Rather than going through the same constructions as in the main text, for the sake of diversity, we will adopt here a complementary approach. We will present a (strictly) positive linear functional on the holonomy C*-algebra $\overline{\mathcal{HA}}$ whose properties suffice to ensure the existence of a diffeomorphism-invariant, faithful, regular measure on the spectrum of $\overline{\mathcal{HA}}$.

We know from the general theory outlined in Section 2 that, to specify a positive linear functional on $\overline{\mathcal{HA}}$, it suffices to provide an appropriate generating functional $\Gamma[\alpha]$ on $\mathcal{L}_{x^0}$. Let us simply set

$$\Gamma(\alpha) = \begin{cases} 1, & \text{if } \tilde{\alpha} = \tilde{o} \\ 0, & \text{otherwise,} \end{cases} \tag{A.6}$$

where $\tilde{o}$ is the identity hoop in $\mathcal{HG}$. It is clear that this functional on $\mathcal{L}_{x^0}$ is diffeomorphism invariant. Indeed, it is the simplest of such functionals on $\mathcal{L}_{x^0}$. We will now show that it does have all the properties to be a generating function[9].

Theorem A.5 *$\Gamma(\alpha)$ extends to a continuous positive linear function on the holonomy C*-algebra $\mathcal{HA}$. Furthermore, this function is strictly positive.*

---

[9]Since this functional is so simple and natural, one might imagine using it to define a positive linear functional also for the $SU(2)$ holonomy algebra of the main text. However, this strategy fails because of the $SU(2)$ identities. For example, in the $SU(2)$ holonomy algebra we have $T_{\alpha \cdot \beta \cdot \beta \cdot \alpha} + T_{\alpha \cdot \beta \cdot \alpha^{-1} \cdot \beta^{-1}} - T_{\alpha \cdot \beta \cdot \alpha \cdot \beta} = 1$. Evaluation of the generating functional $(A.6)$ on this identity would lead to the contradiction $0 = 1$. We *can* define a

Proof By its definition, the loop functional $\Gamma(\alpha)$ admits a well-defined projection on the hoop space $\mathcal{HG}$ which we will denote again by Γ. Let $\mathcal{FHG}$ be the free vector space generated by the hoop group. Extend Γ to $\mathcal{FHG}$ by linearity. We will first establish that this extension satisfies the following property: given a finite set of hoops $\tilde{\alpha}_j, j = 1, \ldots, n$, such that, if $\tilde{\alpha}_i \neq \tilde{\alpha}_j$ if $i \neq j$, we have

$$\Gamma\bigg(\bigg(\sum_{j=1}^{n} a_j \tilde{\alpha}_j\bigg)^\star \bigg(\sum_{j=1}^{n} a_j \tilde{\alpha}_j\bigg)\bigg) < \sup_{A^0 \in \mathcal{A}^0} \bigg(\sum_{j=1}^{n} a_j H(\tilde{\alpha}_j, A^0)\bigg)^\star \bigg(\sum_{j=1}^{n} a_j H(\tilde{\alpha}_j, A^0)\bigg).$$
(A.7)

To see this, note first that, by definition of Γ, the left side equals $\sum |a_j|^2$. The right side equals $(\sum a_j \exp i\theta_j)^\star (\sum a_j \exp i\theta_j)$ where θ_j depend on A^0 and the hoop $\tilde{\alpha}_j$. Now, by Lemma A.3, given the n hoops $\tilde{\alpha}_j$, one can find an A^0 such that the angles θ_j can be made as close as one wishes to a pre-specified n-tuple θ_j^0. Finally, given the n complex numbers a_i, one can always find θ_j^0 such that $\sum |a_j|^2 < |(\sum a_j \exp i\theta_j^0)^\star (\sum a_j \exp i\theta_j^0)|$. Hence we have (A.7).

The inequality (A.7) implies that the functional Γ has a well-defined projection on the holonomy $\star$-algebra $\mathcal{HA}$ (which is obtained by taking the quotient of $\mathcal{FHG}$ by the subspace K consisting of $\sum b_i \tilde{\alpha}_i$ such that $\sum b_i H(\tilde{\alpha}_i, A^0) = 0$ for all $A^0 \in \mathcal{A}^0$). Furthermore, the projection is positive definite; $\Gamma(f^\star f) \geq 0$ for all $f \in \mathcal{HA}$, equality holding *only* if $f = 0$. Next, since the norm on this $\star$-algebra is the sup norm on $\mathcal{A}^0$, it follows from (A.7) that the functional Γ is continuous on $\mathcal{HA}$. Hence it admits a unique continuous extension to the C*-algebra $\overline{\mathcal{HA}}$. Finally, (A.7) implies that the functional continues to be strictly positive on $\overline{\mathcal{HA}}$. □

Theorem A.5 implies that the generating functional Γ provides a continuous, faithful representation of $\overline{\mathcal{HA}}$ and hence a regular, strictly positive measure on the Gel'fand spectrum of this algebra. It is not difficult to verify by direct calculations that this is precisely the $U(1)$ analog of the induced Haar measure discussed in Section 4.

Appendix B: C^ω loops, $U(N)$ and $SU(N)$ connections

In this appendix, we will consider another extension of the results presented in the main text. We will now work with analytic manifolds and piecewise analytic loops as in the main text. However, we will let the manifold have

---

positive linear function on the $SU(2)$ holonomy C*-algebra via

$$\Gamma'(\alpha) = \begin{cases} 1, & \text{if } T_{\tilde{\alpha}}(A) = 1 \, \forall A \in \mathcal{A}^0 \\ 0, & \text{otherwise,} \end{cases}$$

where we have regarded $\mathcal{A}^0$ as a subspace of the space of $SU(2)$ connections. However, on the $SU(2)$ holonomy algebra, this positive linear functional is not strictly positive: the measure on $\overline{\mathcal{A}/\mathcal{G}}$ it defines is concentrated just on $\mathcal{A}^0$. Consequently, the resulting representation of the $SU(2)$ holonomy C*-algebra fails to be faithful.

any dimension and let the gauge group G be either $U(N)$ or $SU(N)$. Consequently, there are two types of complications: algebraic and topological. The first arise because, for example, the products of traces of holonomies can no longer be expressed as sums while the second arise because the connections in question may be defined on non-trivial principal bundles. Nonetheless, we will see that the main results of the paper continue to hold in these cases as well. Several of the results also hold when the gauge group is allowed to be any compact Lie group. However, for brevity of presentation, we will refrain from making digressions to the general case. Also, since most of the arguments are rather similar to those given in the main text, details will be omitted.

Let us begin by fixing a principal fibre bundle $P(\Sigma, G)$, obtain the main results, and then show, at the end, that they are independent of the initial choice of $P(\Sigma, G)$. Definitions of the space $\mathcal{L}_{x^0}$ of based loops, and holonomy maps $H(\alpha, A)$, are the same as in Section 2. To keep the notation simple, we will continue to use the same symbols as in the main text to denote various spaces which, however, now refer to connections on $P(\Sigma, G)$. Thus, $\mathcal{A}$ will denote the space of (suitably regular) connections on $P(\Sigma, G)$, $\mathcal{HG}$ will denote the $P(\Sigma, G)$-hoop group, and $\overline{\mathcal{HA}}$ will be the holonomy C*-algebra generated by (finite sums of finite products of) traces of holonomies of connections in $\mathcal{A}$ around hoops in $\mathcal{HG}$.

B.1 Holonomy C*-algebra

We will set $T_{\tilde{\alpha}}(A) = \frac{1}{N}\mathrm{Tr}H(\alpha, A)$, where the trace is taken in the fundamental representation of $U(N)$ or $SU(N)$, and regard $T_{\tilde{\alpha}}$ as a function on $\mathcal{A}/\mathcal{G}$, the space of gauge-equivalent connections. The holonomy algebra $\mathcal{HA}$ is, by definition, generated by all finite linear combinations (with complex coefficients) of finite products of the $T_{\tilde{\alpha}}$, with the $\star$-relation given by $f^\star = \bar{f}$, where the 'bar' stands for complex conjugation. Being traces of $U(N)$ and $SU(N)$ matrices, $T_{\tilde{\alpha}}$ are bounded functions on $\mathcal{A}/\mathcal{G}$. We can therefore introduce on $\mathcal{HA}$ the sup norm

$$\|f\| = \sup_{A \in \mathcal{A}} |f(A)|,$$

and take the completion of $\mathcal{HA}$ to obtain a C*-algebra. We will denote it by $\overline{\mathcal{HA}}$. This is the holonomy C*-algebra.

The key difference between the structure of this $\mathcal{HA}$ and the one we constructed in Section 2 is the following. In Section 2, the $\star$-algebra $\mathcal{HA}$ was obtained simply by imposing an equivalence relation (K, see eqn (2.7)) on the free vector space generated by the loop space $\mathcal{L}_{x^0}$. This was possible because, owing to $SU(2)$ Mandelstam identities, products of any two $T_{\tilde{\alpha}}$ could be expressed as a linear combination of $T_{\tilde{\alpha}}$ (eqn (2.6)). For the groups under consideration, the situation is more involved. Let us summarize the situation in the slightly more general context of the group $GL(N)$. In this case, the Mandelstam identities follow from the contraction of $N + 1$ matrices—$H(\alpha_1, A), \ldots, H(\alpha_{N+1}, A)$ in our case—with the identity

$$\delta^{j_1}_{[i_1}, \ldots, \delta^{j_{N+1}}_{i_{N+1}]} = 0$$

where δ^j_i is the Kronecker delta and the bracket stands for the antisymmetrization. They enable one to express products of $N + 1$ $T_{\tilde\alpha}$ functions as a linear combination of traces of products of $1, 2, \ldots, N$ $T_{\tilde\alpha}$ functions. Hence, we have to begin by considering the free algebra generated by the hoop group and then impose on it all the suitable identities by extending the definition of the kernel (2.7). Consequently, if the gauge group is $GL(N)$, an element of the holonomy algebra $\mathcal{HA}$ is expressible as an equivalence class $[a\alpha, b\alpha_1 \cdot \alpha_2, \ldots, c\gamma_1 \cdots \gamma_N]$, where a, b, c are complex numbers; products involving $N + 1$ and higher loops are redundant. (For subgroups of $GL(N)$—such as $SU(N)$—further reductions arise.)

Finally, let us note identities that will be useful in what follows. These result from the fact that the determinant of $N \times N$ matrix M can be expressed in terms of traces of powers of that matrix, namely

$$\det M = F(\mathrm{Tr} M, \mathrm{Tr} M^2, \ldots, \mathrm{Tr} M^N)$$

for a certain polynomial F. Hence, if the gauge group G is $U(N)$, we have, for any hoop $\tilde\alpha$,

$$F(T_{\tilde\alpha}, \ldots, T_{\tilde\alpha^N}) F(T_{\tilde\alpha^{-1}}, \ldots, T_{\tilde\alpha^{-N}}) = 1. \tag{B.1a}$$

In the $SU(N)$ case a stronger identity holds:

$$F(T_{\tilde\alpha}, \ldots, T_{\tilde\alpha^N}) = 1. \tag{B.1b}$$

B.2   Loop decomposition and the spectrum theorem

The construction of Section 3.1 for decomposition of a finite number of loops into *independent loops* makes no direct reference to the gauge group and therefore carries over as is. Also, it is still true that the map

$$\mathcal{A} \ni A \mapsto (H(\beta_1, A), \ldots, H(\beta_n, A)) \in G^n$$

is onto if the loops β_i are independent. (The proof requires some minor modifications which consist of using local sections and a sufficiently large number of generators of the gauge group.) A direct consequence is that the analog of Lemma 3.3 continues to hold.

As for the Gel'fand spectrum, the A–I result that there is a natural embedding of $\mathcal{A}/\mathcal{G}$ into the spectrum $\overline{\mathcal{A}/\mathcal{G}}$ of $\overline{\mathcal{HA}}$ goes through as before and so does the general argument due to Rendall [5] that the image of $\mathcal{A}/\mathcal{G}$ is dense in $\overline{\mathcal{A}/\mathcal{G}}$. (This is true for any group and representation for which the traces of holonomies separate the points of $\mathcal{A}/\mathcal{G}$.) Finally, using the analog of Lemma 3.3, it is straightforward to establish the analog of the main conclusion of Lemma 3.4.

The converse—the analog of Lemma 3.2—on the other hand requires further analysis based on Giles' results along the lines used by A–I in the $SU(2)$ case. For the gauge group under consideration, given an element of the spectrum $\bar{A}$—i.e. a continuous homomorphism from $\overline{\mathcal{H}\mathcal{A}}$ to the $\star$-algebra of complex numbers—Giles' theorem [10] provides us with a homomorphism $\hat{H}_{\bar{A}}$ from the hoop group $\mathcal{H}\mathcal{G}$ to $GL(N)$.

What we need to show is that $H_{\bar{A}}$ can be so chosen that it takes values in the given gauge group G. We can establish this by considering the eigenvalues $\lambda_1, \ldots, \lambda_N$ of $H_{\bar{A}}(\tilde{\alpha})$. (After all, the characteristic polynomial of the matrix $H_{\bar{A}}(\tilde{\alpha})$ is expressible directly by the values of $\bar{A}$ taken on the hoops $\tilde{\alpha}, \tilde{\alpha}^2, \ldots, \tilde{\alpha}^n$.) First, we note from eqn (B.1a) that in particular $\lambda_i \neq 0$, for every i. Thus far, we have used only the algebraic properties of $\bar{A}$. From continuity it follows that for every hoop $\tilde{\alpha}$, $\mathrm{Tr}H_{\bar{A}}(\tilde{\alpha}) \leq N$. Substituting for $\tilde{\alpha}$ its powers we conclude that

$$\|\lambda_1^k + \cdots + \lambda_N^k\| \leq N$$

for every integer k. This suffices to conclude that

$$\|\lambda_i\| = 1.$$

With this result in hand, we can now use Giles' analysis to conclude that given $\bar{A}$ we can find $H_{\bar{A}}$ which takes values in $U(N)$. If G is $SU(N)$ then, by the identity $(B.1b)$, we have

$$\det H_{\bar{A}} = 1,$$

so that the analog of Lemma 3.2 holds.

Combining these results, we have the spectrum theorem: every element $\bar{A}$ of the spectrum $\overline{\mathcal{A}/\mathcal{G}}$ defines a homomorphism $\hat{H}$ from the hoop group $\mathcal{H}\mathcal{G}$ to the gauge group G, and, conversely, every homomorphism $\hat{H}$ from the hoop group $\mathcal{H}\mathcal{G}$ to G defines an element $\bar{A}$ of the spectrum $\overline{\mathcal{A}/\mathcal{G}}$ such that $\bar{A}(\tilde{\alpha}) = \frac{1}{N}\mathrm{Tr}\hat{H}(\tilde{\alpha})$ for all $\tilde{\alpha} \in \mathcal{H}\mathcal{G}$. Two homomorphisms $\hat{H}_1$ and $\hat{H}_2$ define the same element of the spectrum if and only if $\mathrm{Tr}\hat{H}_2 = \mathrm{Tr}\hat{H}_1$.

B.3 Cylindrical functions and the induced Haar measure

Using the spectrum theorem, we can again associate with any subgroup $\mathcal{S}^\star \equiv \mathcal{H}\mathcal{G}(\tilde{\beta}_1, \ldots, \tilde{\beta}_n)$ generated by n independent hoops $\tilde{\beta}_1, \ldots, \tilde{\beta}_n$, an equivalence relation $\sim$ on $\overline{\mathcal{A}/\mathcal{G}}$: $\bar{A}_1 \sim \bar{A}_2$ iff $\bar{A}_1(\tilde{\gamma}) = \bar{A}_2(\tilde{\gamma})$ for all $\tilde{\gamma} \in \mathcal{S}^\star$. This relation provides us with a family of projections from $\overline{\mathcal{A}/\mathcal{G}}$ onto the compact manifolds G^n/Ad (since, for groups under consideration, the traces of all elements of a finitely generated subgroup in the fundamental representation suffice to characterize the subgroup modulo an overall adjoint map). Therefore, as before, we can define cylindrical functions on $\overline{\mathcal{A}/\mathcal{G}}$ as the pullbacks to $\overline{\mathcal{A}/\mathcal{G}}$ of continuous functions on G^n/Ad. They again form a C*-algebra. Using the fact that traces of products of group elements suffice to separate points of G^n/Ad when $G = U(N)$ or $G = SU(N)$ [16], it again follows that the C*-algebra of cylindrical functions is isomorphic with $\overline{\mathcal{HA}}$. Finally, the construction of Section 4.2 which led us to the definition of the induced Haar measure goes through step by step. The resulting representation of the C*-algebra $\overline{\mathcal{HA}}$ is again faithful and diffeomorphism invariant.

B.4 Bundle dependence

In all the constructions above, we fixed a principal bundle $P(\Sigma, G)$. To conclude, we will show that the C*-algebra $\overline{\mathcal{HA}}$ and hence its spectrum are *independent* of this choice. First, as noted at the end of Section 3.1, using the hoop decomposition one can show that two loops in $\mathcal{L}_{x^0}$ define the same hoop if and only if they differ by a combination of reparametrization and (an immediate) retracing of segments for gauge groups under consideration. Therefore, the hoop groups obtained from any two bundles are the same.

Fix a gauge group G from $\{U(N), SU(N)\}$ and let $P_1(\Sigma, G)$ and $P_2(\Sigma, G)$ be two principal bundles. Let the corresponding holonomy C*-algebras be $\overline{\mathcal{HA}}^{(1)}$ and $\overline{\mathcal{HA}}^{(2)}$. Using the spectrum theorem, we know that their spectra are naturally isomorphic: $\overline{\mathcal{A}/\mathcal{G}}^{(1)} = \mathrm{Hom}(\mathcal{HG}, G) = \overline{\mathcal{A}/\mathcal{G}}^{(2)}$. We wish to show that the algebras are themselves naturally isomorphic, i.e. that the map $\mathcal{I}$ defined by

$$\mathcal{I} \circ \left(\sum_{i=1}^{n} a_i T_{\tilde{\alpha}_i}^{(1)} \right) = \sum_{i=1}^{n} a_i T_{\tilde{\alpha}_i}^{(2)} \tag{B.2}$$

is an isomorphism from $\overline{\mathcal{HA}}^{(1)}$ to $\overline{\mathcal{HA}}^{(2)}$. Let us decompose the given hoops $\tilde{\alpha}_1, \ldots, \tilde{\alpha}_k$ into independent hoops $\tilde{\beta}_1, \ldots, \tilde{\beta}_n$. Let $\mathcal{S}^\star$ be the subgroup of the hoop group they generate and, as in Section 4.1, let $\pi^{(i)}(\mathcal{S}^\star)$, $i = 1, 2$, be the corresponding projection maps from $\overline{\mathcal{A}/\mathcal{G}}$ to G^n/Ad. From the definition of the maps it follows that the projections of the two functions in $(B.2)$ to G^N/Ad are in fact equal. Hence, we have

$$\| \sum_{i=1}^{n} a_i T_{\tilde{\alpha}_i}^{(1)} \|_1 = \| \sum_{i=1}^{n} a_i T_{\tilde{\alpha}_i}^{(2)} \|_2,$$

whence it follows that $\mathcal{I}$ is an isomorphism of C*-algebras. Thus, it does not matter which bundle we begin with; we obtain the same holonomy C*-algebra $\overline{\mathcal{HA}}$. Now, fix a bundle, P_1 say, and consider a connection A_2 defined on P_2. Note that the holonomy map of A_2 defines a homomorphism from the (bundle-independent) hoop group to G. Hence, A_2 defines also a point in the spectrum of $\overline{\mathcal{HA}}^{(1)}$. Therefore, *given a manifold M and a gauge group G the spectrum $\overline{\mathcal{A}/\mathcal{G}}$ automatically contains all the connections* *on all the principal G-bundles over M.*

Acknowledgements

The authors are grateful to John Baez for several illuminating comments and to Roger Penrose for suggesting that the piecewise analytic loops may be better suited for their purpose than the piecewise smooth ones. JL also thanks Clifford Taubes for stimulating discussions. This work was supported in part by the National Science Foundation grants PHY93–96246 and PHY91–07007; by the Polish Committee for Scientific Research (KBN) through grant no. 2 0430 9101; and by the Eberly research fund of Pennsylvania State University.

Bibliography

1. A. Ashtekar, *Non-perturbative canonical quantum gravity* (Notes prepared in collaboration with R.S. Tate), World Scientific, Singapore (1991)

2. A. Ashtekar and C. Isham, *Classical & Quantum Gravity* **9**, 1433–1485 (1992)

3. R. Gambini and A. Trias, *Nucl. Phys.* **B278**, 436–448 (1986)

4. C. Rovelli and L. Smolin, *Nucl. Phys.* **B331**, 80–152 (1992)

5. A. Rendall, *Classical & Quantum Gravity* **10**, (1993) 605–608

6. J. Baez, In: *Proceedings of the Conference on Quantum Topology*, eds L. Crane and D. Yetter (to appear), hep–th/9305045

7. J. Lewandowski, *Classical & Quantum Gravity* **10**, 879–904 (1993)

8. P.K. Mitter and C. Viallet, *Commun. Math. Phys.* **79**, 43–58 (1981)

9. C. Di Bartolo, R. Gambini, and J. Griego, preprint IFFI/9301; gr–qc/9303010

10. R. Giles, *Phys. Rev.* **D24**, 2160–2168 (1981)

11. J.W. Barrett, *Int. J. Theor. Phys.* **30**, 1171–1215 (1991)

12. A.N. Kolmogorov, *Foundations of the Theory of Probability*, Chelsea, New York (1956)

13. Y. Choquet-Bruhat, C. DeWitt-Morette and M. Dillard-Bleick, *Analysis, Manifolds and Physics*, North-Holland, Amsterdam (1982), pp. 573–587

14. M. Creutz, *Quarks, Gluons and Lattices*, Cambridge University Press, Cambridge (1988) Chapter 8

15. A. Trautman, *Czech. J. Phys.* **B29**, 107–111 (1979)

16. C. Procesi, *Adv. Math.* **19**, 306–381 (1976)

The Gauss Linking Number in Quantum Gravity

Rodolfo Gambini

Intituto de Física, Facultad de Ciencias,
Tristan Narvaja 1674, Montevideo, Uruguay
(email: rgambini@fising.edu.uy)

Jorge Pullin

Center for Gravitational Physics and Geometry,
Pennsylvania State University,
University Park, Pennsylvania 16802, USA
(email: pullin@phys.psu.edu)

Abstract

We show that the exponential of the Gauss (self-)linking number of a knot is a solution of the Wheeler–DeWitt equation in loop space with a cosmological constant. Using this fact, it is straightforward to prove that the second coefficient of the Jones polynomial is a solution of the Wheeler–DeWitt equation in loop space with no cosmological constant. We perform calculations from scratch, starting from the connection representation and give details of the proof. Implications for the possibility of generation of other solutions are also discussed.

1 Introduction

The introduction of the loop representation for quantum gravity has made it possible for the first time to find solutions to the Wheeler–DeWitt equation (the quantum Hamiltonian constraint) and therefore to have possible candidates to become physical wave functions of the gravitational field. In the loop representation the Hamiltonian constraint has non-vanishing action only on functions of intersecting loops. It was first argued that by considering wave functions with support on smooth loops one could solve the constraint straightforwardly [1, 2]. However, it was later realized that such solutions are associated with degenerate metrics (metrics with zero determinant) and this posed inconsistencies if one wanted to couple the theory [3]. For instance, if one considered general relativity with a cosmological constant it turns out that non-intersecting loop states also solve the Wheeler–DeWitt equation for arbitrary values of the cosmological constant. This does not appear as reasonable since different values of the cosmological constant lead to widely different behaviors in general relativity.

Therefore the problem of solving the Wheeler–DeWitt equation in loop

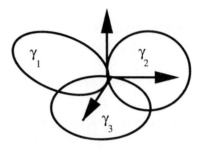

FIG. 1. At least a triple intersection is needed to have a state associated
with a non-degenerate 3-dimensional metric in the loop representation.
And even in this case the metric is non-degenerate only at the point of
intersection.

space is far from solved and has to be tackled by considering the action of
the Hamiltonian constraint in the loop representation on states based on
(at least triply) intersecting loops, as depicted in Fig. 1.

Although performing this kind of direct calculations is now possible,
since well-defined expressions for the Hamiltonian constraint exist in terms
of the loop derivative [4] (see also [3]) and actually some solutions have been
found with this approach [5], another line of reasoning has also proved to
be useful.

This other approach is based on the fact that the exponential of the
Chern–Simons term based on Ashtekar's connection

$$\Psi^{CS}_\Lambda[A] = \exp\left(-\tfrac{6}{\Lambda}\int d^3x (A^i_a \partial_b A^i_c + \tfrac{2}{3} A^i_a A^j_b A^k_c \epsilon^{ijk})\epsilon^{abc}\right) \tag{1.1}$$

is a solution of the Wheeler–DeWitt equation in the connection representation with a cosmological term

$$\hat{H} = \epsilon^{ijk}\frac{\delta}{\delta A^i_a}\frac{\delta}{\delta A^j_b}F^k_{ab} + \tfrac{\Lambda}{6}\epsilon^{ijk}\epsilon^{abc}\frac{\delta}{\delta A^i_a}\frac{\delta}{\delta A^j_b}\frac{\delta}{\delta A^k_c} \tag{1.2}$$

(the first term is just the usual Hamiltonian constraint with $\Lambda = 0$ and the
second term is just $\Lambda \det g$ where $\det g$ is the determinant of the spatial
part of the metric written in terms of triads).

To prove this fact one simply needs to notice that for the Chern–Simons
state introduced above,

$$\frac{\delta}{\delta A^i_a}\Psi^{CS}[A] = -\tfrac{6}{\Lambda}\epsilon^{abc}F^i_{bc}\Psi^{CS}[A], \tag{1.3}$$

and therefore the rightmost functional derivative in the cosmological constant term of the Hamiltonian (1.2) produces a term in F^i_{ab} that exactly
cancels the contribution from the vacuum Hamiltonian constraint.

This result for the Chern–Simons state in the connection representation has an immediate counterpart in the loop representation, since the transform of the Chern–Simons state into the loop representation

$$\Psi_\Lambda^{CS}(\gamma) = \int dA \, \Psi_\Lambda^{CS}[A] \, W_\gamma[A] \qquad (1.4)$$

can be interpreted as the expectation value of the Wilson loop in a Chern–Simons theory,

$$\Psi_\Lambda^{CS}(\gamma) = \int dA \, e^{-\frac{6}{\Lambda} S_{CS}} \, W_\gamma[A] = \langle W_\gamma \rangle \qquad (1.5)$$

with coupling constant $\frac{6}{\Lambda}$, and we know due to the insight of Witten [6] that this coincides with the Kauffman bracket of the loop. Therefore the state

$$\Psi^{CS}(\gamma) = \text{Kauffman bracket}_\Lambda(\gamma) \qquad (1.6)$$

should be a solution of the Wheeler–DeWitt equation in loop space.

This last fact can actually be checked in a direct fashion using the expressions for the Hamiltonian constraint in loop space of [4]. In order to do this we make use of the identity

$$\text{Kauffman bracket}_\Lambda(\gamma) = e^{\Lambda \text{Gauss}(\gamma)} \text{Jones polynomial}_\Lambda(\gamma) \qquad (1.7)$$

which relates the Kauffman bracket, the Gauss (self-)linking number, and the Jones polynomial. The Gauss self-linking number is framing dependent, and so is the Kauffman bracket. The Jones polynomial, however, is framing independent. This raises the issue of up to what extent statements about the Kauffman bracket being a state of gravity are valid, since one expects states of quantum gravity to be truly diffeomorphism-invariant objects and framing is always dependent on an external device which should conflict with the diffeomorphism invariance of the theory. Unfortunately it is not clear at present how to settle this issue since it is tied to the regularization procedures used to define the constraints. We will return to these issues in the final discussion.

We now expand both the Jones polynomial and the exponential of the Gauss linking number in terms of Λ and get the expression

$$\begin{aligned}
\text{Kauffman bracket}_\Lambda(\gamma) \;=\;\; & 1 + \text{Gauss}(\gamma)\Lambda \\
+ \;\; & (\text{Gauss}(\gamma)^2 + a_2(\gamma))\Lambda^2 \\
+ \;\; & (\text{Gauss}(\gamma)^3 + \text{Gauss}(\gamma)^2 a_2(\gamma) + a_3(\gamma))\Lambda^3 \\
+ \;\; & \cdots
\end{aligned} \qquad (1.8)$$

where a_2, a_3 are the second and third coefficient of the infinite expansion of the Jones polynomial in terms of Λ. a_2 is known to coincide with the second coefficient of the Conway polynomial [7].

Rodolfo Gambini and Jorge Pullin

One could now apply the Hamiltonian constraint in the loop representation with a cosmological constant to the expansion (1.8) and one would find that certain conditions have to be satisfied if the Kauffman bracket is to be a solution. Among them it was noticed [8] that

$$\hat{H}_0 a_2(\gamma) = 0 \tag{1.9}$$

where $\hat{H}_0$ is the Hamiltonian constraint without cosmological constant (more recently it has also been shown that $\hat{H}_0 a_3(\gamma) = 0$ [9] but we will not discuss it here).

Summarizing, the fact that the Kauffman bracket is a solution of the Wheeler–DeWitt equation with cosmological constant seems to have as a direct consequence that the coefficients of the Jones polynomial are solutions of the vacuum ($\Lambda = 0$) Wheeler–DeWitt equation! This partially answers the problem of framing we pointed out above. Even if one is reluctant to accept the Kauffman bracket as a state because of its framing dependence, it can be viewed as an intermediate step of a framing-dependent proof that the Jones polynomial (which is a framing-independent invariant) solves the Wheeler–DeWitt equation (it should be stressed that we only have evidence that the first two non-trivial coefficients are solutions).

The purpose of this paper is to present a rederivation of these facts from a different, and to our understanding simpler, perspective. We will show that the (framing-dependent) knot invariant

$$\Psi_\Lambda^G(\gamma) = e^{\Lambda \text{Gauss}(\gamma)} \tag{1.10}$$

is also a solution of the Wheeler–DeWitt equation with cosmological constant. It can be viewed as an 'Abelian limit' of the Kauffman bracket (more on this in the conclusions). Given this fact, one can therefore consider their difference divided by Λ^2,

$$D_\Lambda(\gamma) = \frac{(\text{Kauffman bracket}_\Lambda(\gamma) - \Psi_\Lambda^G(\gamma))}{\Lambda^2}, \tag{1.11}$$

which is also a solution of the Wheeler–DeWitt equation with a cosmological constant. This difference is of the form

$$D_\Lambda(\gamma) = a_2(\gamma) + (a_3(\gamma) + \text{Gauss}(\gamma)^2 a_2)\Lambda + \dots. \tag{1.12}$$

Now, this difference is a state for all values of Λ, in particular for $\Lambda = 0$. This means that $a_2(\gamma)$ should be a solution of the Wheeler–DeWitt equation. This confirms the proof given in [5, 8].

Therefore we see that by noticing that the Gauss linking number is a state with cosmological constant, it is easy to prove that the second coefficient of the infinite expansion of the Jones polynomial (which coincides with the second coefficient of the Conway polynomial) is a solution of the Wheeler–DeWitt equation with $\Lambda = 0$.

The rest of this paper will be devoted to a detailed proof that the exponential of the Gauss linking number solves the Wheeler–DeWitt equation with cosmological constant. To this aim we will derive expressions for the Hamiltonian constraint with cosmological constant in the loop representation. We will perform the calculation explicitly for the case of a triply self-intersecting loop, the more interesting case for gravity purposes (it should be noticed that all the arguments presented above were independent of the number and order of intersections of the loops; we just present the explicit proof for a triple intersection since in three spatial dimensions it represents the most important type of intersection).

Apart from presenting this new state, we think the calculations exhibited in this paper should help the reader get into the details of how these calculations are performed and make an intuitive contact between the expressions in the connection and the loop representation.

In Section 2 we derive the expression of the Hamiltonian constraint (with a cosmological constant) in the loop representation for a triply intersecting loop in terms of the loop derivative. In Section 3 we write an explicit analytic expression for the Gauss linking number and prove that it is a state of the theory. We end in Section 4 with a discussion of the results.

2 The Wheeler–DeWitt equation in terms of loops

Here we derive the explicit form in the loop representation of the Hamiltonian constraint with a cosmological constant. The derivation proceeds along the following lines. Suppose one wants to define the action of an operator $\hat{O}_L$ on a wave function in the loop representation $\Psi(\gamma)$. Applying the transform

$$\hat{O}_L \Psi(\gamma) \equiv \int dA \hat{O}_L W_\gamma[A] \Psi[A] \qquad (2.1)$$

the operator $\hat{O}$ in the right member acts on the loop dependence of the Wilson loop. On the other hand, this definition should agree with

$$\hat{O}_L = \int dA W_\gamma[A] \hat{O}_C \Psi[A] \qquad (2.2)$$

where $\hat{O}_C$ is the connection representation version of the operator in question. Therefore, it is clear that

$$\hat{O}_L W_\gamma[A] \equiv \hat{O}_C^\dagger W_\gamma[A], \qquad (2.3)$$

where † means the adjoint operator with respect to the measure of integration dA. If one assumes that the measure is trivial, the only effect of taking the adjoint is to reverse the factor ordering of the operators.

Concretely, in the case of the Hamiltonian constraint (without cosmological constant)

$$\hat{H}_C = \epsilon^{ijk} \frac{\delta}{\delta A_a^i} \frac{\delta}{\delta A_b^j} F_{ab}^k \tag{2.4}$$

and therefore

$$\hat{H}_L W_\gamma[A] \equiv \epsilon^{ijk} F_{ab}^k \frac{\delta}{\delta A_a^i} \frac{\delta}{\delta A_b^j} W_\gamma[A]. \tag{2.5}$$

We now need to compute this quantity explicitly. For that we need the expression of the functional derivative of the Wilson loop with respect to the connection

$$\frac{\delta}{\delta A_a^i(x)} W_\gamma[A] = \oint dy^a \delta(y - x) \text{Tr}\left[\text{Pexp}\left(\int_o^y dz^b A_b\right) \tau^i \text{Pexp}\left(\int_y^o dz^b A_b\right)\right] \tag{2.6}$$

where o is the basepoint of the loop. The Wilson loop is therefore 'broken' at the point of action of the functional derivative and a Pauli matrix (τ^i) is inserted. It is evident that with this action of the functional derivative the Hamiltonian operator is not well defined. We need to regularize it as

$$\hat{H}_L W_\gamma[A] = \lim_{\epsilon \to 0} f_\epsilon(x - z)\epsilon^{ijk} F_{ab}^k(x) \frac{\delta}{\delta A_a^i(x)} \frac{\delta}{\delta A_b^j(z)} W_\gamma[A] \tag{2.7}$$

where $f_\epsilon(x - z) \to \delta(x - z)$ when $\epsilon \to 0$. Caution should be exercised, since such point splitting breaks the gauge invariance of the Hamiltonian. There are a number of ways of fixing this situation in the language of loops. One of them is to define the Hamiltonian inserting pieces of holonomies connecting the points x and z between the functional derivatives to produce a gauge-invariant quantity [1]. Here we will only study the operator in the limit in which the regulator is removed; therefore we will not be concerned with these issues. A proper calculation would require their careful study.

It is immediate from the above definitions that the Hamiltonian constraint vanishes in any regular point of the loop, since it yields a term $dy^a dy^b F_{ab}^i$ which vanishes due to the antisymmetry of F_{ab}^i and the symmetry of $dy^a dy^b$ at points where the loop is smooth. However, at intersections there can be non-trivial contributions. Here we compute the contribution at a point of triple self-intersection,

$$\epsilon^{ijk} F_{ab}^k(x) \frac{\delta}{\delta A_a^i(x)} \frac{\delta}{\delta A_b^j(z)} W_{\gamma_1 \circ \gamma_2 \circ \gamma_3}[A] = \epsilon^{ijk} F_{ab}^k(x) \tag{2.8}$$

$$\times (\dot{\gamma}_1^a \dot{\gamma}_2^b W_{\gamma_1 \tau^j \gamma_2 \tau^k \gamma_3}[A] + \dot{\gamma}_1^a \dot{\gamma}_3^b W_{\gamma_1 \tau^j \gamma_2 \gamma_3 \tau^k}[A] + \dot{\gamma}_2^a \dot{\gamma}_3^b W_{\gamma_1 \gamma_2 \tau^j \gamma_3 \tau^k}[A]),$$

where we have denoted $\gamma = \gamma_1 \circ \gamma_2 \circ \gamma_3$ where γ_i are the 'petals' of the loop as indicated in Fig. 1. By $W_{\gamma_1 \tau^j \gamma_2 \tau^k \gamma_3}[A]$ we really mean take the holonomy from the basepoint along γ_1 up to just before the intersection point, insert a Pauli matrix, continue along γ_1 to the intersection, continue along

γ_2, and just before the intersection insert another Pauli matrix, continue to the intersection, and complete the loop along γ_3. One could pick 'after' the intersection instead of 'before' to include the Pauli matrices and it would make no difference since we are concentrating on the limit in which the regulator is removed, in which the insertions are done at the intersection. By $\dot{\gamma}_1^a$ we mean the tangent to the petal number 1 just before the intersection (where the Pauli matrix was inserted). This is just shorthand for expressions like $\oint dy^a \delta(x-y)$ when the point x is close to the intersection, so strictly speaking $\dot{\gamma}_1^a$ really is a distribution that is non-vanishing only at the point of intersection.

One now uses the following identity for traces of $SU(2)$ matrices,

$$\epsilon^{ijk} W_{\alpha\tau^j\beta\tau^k}[A] = \tfrac{1}{2}(W_{\alpha\tau^i}[A]W_\beta[A] - W_\alpha[A]W_{\beta\tau^i}[A]), \qquad (2.9)$$

which is a natural generalization to the case of loops with insertions of the $SU(2)$ Mandelstam identities,

$$W_\alpha[A]W_\beta[A] = W_{\alpha\beta}[A] + W_{\alpha\bar{\beta}}[A], \qquad (2.10)$$

where $\bar{\beta}$ means the loop opposite to β. The result of the application of these identities to the expression of the Hamiltonian is

$$\hat{H}W_\gamma[A] = \qquad\qquad\qquad\qquad\qquad\qquad\qquad\qquad (2.11)$$
$$\tfrac{1}{2}F_{ab}^i \left(\dot{\gamma}_1^a \dot{\gamma}_2^b W_{\bar{\gamma}_2\gamma_3\gamma_1\tau^i}[A] - \dot{\gamma}_1^a \dot{\gamma}_3^b W_{\bar{\gamma}_1\gamma_2\gamma_3\tau^i}[A] + \dot{\gamma}_1^a \dot{\gamma}_2^b W_{\bar{\gamma}_3\gamma_1\gamma_2\tau^i}[A] \right).$$

This expression can be further rearranged making use of the loop derivative. The loop derivative $\Delta_{ab}(\pi_o^x)$ [10, 4] is the differentiation operator that appears in loop space when one considers two loops to be 'close' if they differ by an infinitesimal loop appended through a path π_o^x going from the basepoint to a point of the manifold x as shown in Fig. 2. Its definition is

$$\Psi(\pi_o^x \delta\gamma \pi_x^o \gamma) = (1 + \sigma^{ab}\Delta_{ab}(\pi_o^x))\Psi(\gamma) \qquad (2.12)$$

where σ^{ab} is the element area of the infinitesimal loop $\delta\gamma$ and by $\pi_o^x \delta\gamma \pi_x^o \gamma$ we mean the loop obtained by traversing the path π from the basepoint to x, the infinitesimal loop $\delta\gamma$, the path π from x to the basepoint, and then the loop γ.

We will not discuss all its properties here. The only one we need is that the loop derivative of a Wilson loop taken with a path along the loop is given by

$$\Delta_{ab}(\gamma_o^x)W_\gamma[A] = \mathrm{Tr}\left[F_{ab}(x)\mathrm{Pexp}\left(\oint dy^c A_c \right) \right], \qquad (2.13)$$

which reflects the intuitive notion that a holonomy of an infinitesimal loop is related to the field tensor. Therefore we can write expressions like

$$F_{ab}^i W_{\bar{\gamma}_2\gamma_3\gamma_1\tau^i}[A] \qquad (2.14)$$

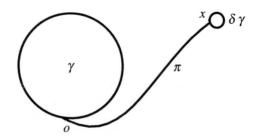

FIG. 2. The loop defining the loop derivative

as

$$\Delta_{ab}(\gamma_1) W_{\bar{\gamma}_2\gamma_3\gamma_1}[A], \qquad (2.15)$$

and the final expression for the Hamiltonian constraint in the loop representation can therefore be read off as follows:

$$\hat{H}\Psi(\gamma) = \tfrac{1}{2}\big[\dot{\gamma}_1^a \dot{\gamma}_2^b \Delta_{ab}(\gamma_1)\Psi(\bar{\gamma}_2\gamma_3\gamma_1)$$
$$+\dot{\gamma}_1^a \dot{\gamma}_3^b \Delta_{ab}(\gamma_3)\Psi(\bar{\gamma}_1\gamma_2\gamma_3) + \dot{\gamma}_1^a \dot{\gamma}_2^b \Delta_{ab}(\gamma_3)\Psi(\bar{\gamma}_3\gamma_1\gamma_2)\big]. \quad (2.16)$$

This expression could be obtained by particularizing that of [4] to the case of a triply self-intersecting loop and rearranging terms slightly using the Mandelstam identities. However, we thought that a direct derivation for this particular case would be useful for pedagogical purposes.

We now have to find the loop representation form of the operator corresponding to the determinant of the metric in order to represent the second term in (1.2). We proceed in a similar fashion, first computing the action of the operator in the connection representation

$$\widehat{\det q} = \epsilon^{ijk}\epsilon_{abc}\frac{\delta}{\delta A_a^i}\frac{\delta}{\delta A_b^j}\frac{\delta}{\delta A_c^k} \qquad (2.17)$$

on a Wilson loop

$$\epsilon^{ijk}\epsilon_{abc}\frac{\delta}{\delta A_a^i}\frac{\delta}{\delta A_b^j}\frac{\delta}{\delta A_c^k}W_\gamma[A] = \epsilon_{abc}\epsilon^{ijk}\dot{\gamma}_1^a\dot{\gamma}_2^b\dot{\gamma}_2^c\left(W_{\gamma_1\tau^i\gamma_2\tau^j\gamma_3\tau^k}[A]\right). \quad (2.18)$$

This latter expression can be rearranged with the following identity between holonomies with insertions of Pauli matrices:

$$\epsilon^{ijk}W_{\gamma_1\tau^i\gamma_2\tau^j\gamma_3\tau^k}[A] = \tfrac{1}{4}W_{\gamma_1\gamma_3\bar{\gamma}_2}[A] + W_{\gamma_2\gamma_1\bar{\gamma}_3}[A] + W_{\gamma_2\gamma_3\bar{\gamma}_1}[A]. \quad (2.19)$$

It is therefore immediate to find the expression of the determinant of the metric in the loop representation:

$$\hat{\det q}\Psi(\gamma) = -\tfrac{1}{4}\epsilon_{abc}\dot{\gamma}_1^a\dot{\gamma}_2^b\dot{\gamma}_3^c\big[\Psi(\gamma_1\gamma_3\bar{\gamma}_2) + \Psi(\gamma_2\gamma_1\bar{\gamma}_3) + \Psi(\gamma_2\gamma_3\bar{\gamma}_1)\big]. \quad (2.20)$$

With these elements we are in a position to perform the main calculation of this paper, to show that the exponential of the Gauss self-linking number of a loop is a solution of the Hamiltonian constraint with a cosmological constant.

3 The Gauss (self-)linking number as a solution

In order to be able to apply the expressions we derived in the previous section for the constraints to the Gauss self-linking number we need an expression for it in terms of which it is possible to compute the loop derivative. This is furnished by the well-known integral expression

$$\text{Gauss}(\gamma) = \frac{1}{4\pi} \oint_\gamma dx^a \oint_\gamma dy^b \epsilon_{abc} \frac{(x-y)^c}{|x-y|^3} \tag{3.1}$$

where $|x-y|$ is the distance between x and y with a fiducial metric. This formula is most well known when the two loop integrals are computed along different loops. In that case the formula gives an integer measuring the extent to which the loops are linked. In the present case we are considering the expression of the linking of a curve with itself. This is in general not well defined without the introduction of a framing [7].

We will rewrite the above expression in a more convenient fashion:

$$\text{Gauss}(\gamma) = \int d^3x \int d^3y X^a(x,\gamma) X^b(y,\gamma) g_{ab}(x,y) \tag{3.2}$$

where the vector densities X are defined as

$$X^a(x,\gamma) = \oint_\gamma dz^a \delta(z-x) \tag{3.3}$$

and the quantity $g_{ab}(x,y)$ is the propagator of a Chern–Simons theory [7],

$$g_{ab}(x,y) = \epsilon_{abc} \frac{(x-y)^c}{|x-y|^3}. \tag{3.4}$$

For calculational convenience it is useful to introduce the notation

$$\text{Gauss}(\gamma) = X^{ax}(\gamma) X^{by}(\gamma) g_{ax\,by} \tag{3.5}$$

where we have promoted the point dependence in x, y to a 'continuous index' and assumed a 'generalized Einstein convention' which means sum over repeated indices a, b and integrate over the 3-manifold for repeated continuous indices. This notation is also faithful to the fact that the index a behaves as a vector density index at the point x; that is, it is natural to pair a and x together.

The only dependence on the loop of the Gauss self-linking number is through the X, so we just need to compute the action of the loop derivative on one of them to be in a position to perform the calculation straightfor-

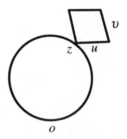

FIG. 3. The loop derivative that appears in the definition of the Hamiltonian is evaluated along a path that follows the loop

wardly. In order to do this we apply the definition of loop derivative, that is we consider the change in the X when one appends an infinitesimal loop to the loop γ as illustrated in Fig. 3. We partition the integral in a portion going from the basepoint to the point z where we append the infinitesimal loop, which we characterize as four segments along the integral curves of two vector fields u^a and v^b of associated lengths ϵ_1 and ϵ_2, and then we continue from there back to the basepoint along the loop. Therefore,

$$\left((1 + \sigma^{ab}\Delta_{ab}(\gamma_o^z))\right) X^{ax}(\gamma) \equiv \int_{\gamma_o^z} dy^a \delta(x - y)\epsilon_1 u^a \delta(x - z)$$

$$+ \epsilon_2 v^b(1 + u^c \partial_c)\delta(x - z) - \epsilon_1 u^a(1 + [u^c + v^c]\partial_c)\delta(x - z) \quad (3.6)$$

$$- \epsilon_2 v^a(1 + v^c \partial_c)\delta(x - z) + \int_{\gamma_z^o} dy^a \delta(y - z).$$

The last and first term combine to give back $X^{ax}(\gamma)$ and therefore one can read off the action of the loop derivative from the other terms. Rearranging one gets

$$\Delta_{ab}(\gamma_o^z)X^{cx}(\gamma) = \partial_{[a}\delta_{b]}^c(x - z) \quad (3.7)$$

where the notation $\delta_b^c(x - z)$ stands for $\delta_b^c\delta(x - z)$, the product of the Kronecker and Dirac deltas.

This is really all we need to compute the action of the Hamiltonian.

We therefore now consider the action of the vacuum ($\Lambda = 0$) part of the Hamiltonian on the exponential of the Gauss linking number. The action of the loop derivative is

$$\Delta_{ab}(\gamma_o^x) \exp\left(X^{cy}(\gamma)X^{dz}(\gamma)g_{cy\,dz}\right) = \quad (3.8)$$

$$2\partial_{[a}\delta_{b]}^c(x - y)X^{dz}(\gamma)g_{cy\,dz} \exp\left(X^{ew}(\gamma)X^{fw'}(\gamma)g_{ew\,fw'}\right).$$

Now we must integrate by parts. Using the fact that $\partial_a X^{ax}(\gamma) = 0$ and the definition of $g_{cy\ dz}$ (3.4), we get

$$\Delta_{ab}(\gamma_o^x) \exp\left(X^{cy}(\gamma)X^{dz}(\gamma)g_{cy\ dz}\right) = \tag{3.9}$$
$$2\epsilon_{abc}\left(X^{cx}(\gamma_1) + X^{cx}(\gamma_2) + X^{cx}(\gamma_3)\right) \exp\left(X^{cy}(\gamma)X^{dz}(\gamma)g_{cy\ dz}\right).$$

Therefore the action of the Hamiltonian constraint on the Gauss linking number is

$$\hat{H}e^{\Lambda \text{Gauss}} = \epsilon_{abc}\dot\gamma_1^a\dot\gamma_2^b\dot\gamma_3^c e^{\Lambda \text{Gauss}} \tag{3.10}$$

where we again have replaced the distributional tangents at the point of intersection by an expression only involving the tangents. The expression is only formal since in order to do this a divergent factor should be kept in front. We assume such factors coming from all terms to be similar and therefore ignore them.

It is straightforward now to check that applying the determinant of the metric one the Gauss linking number one gets a contribution exactly equal and opposite by inspection from expression (2.20). This concludes the main proof of this paper.

4 Discussion

We showed that the exponential of the Gauss (self-)linking number is a solution of the Hamiltonian constraint of quantum gravity with a cosmological constant. This naturally can be viewed as the 'Abelian' limit of the solution given by the Kauffman bracket.

What about the issue of regularization? The proof we presented is only valid in the limit where $\epsilon \to 0$, that is, when the regulator is removed. If one does not take the limit the various terms do not cancel. However, the expression for the Hamiltonian constraint we introduced is also only valid when the regulator is removed. A regularized form of the Hamiltonian constraint in the loop representation is more complicated than the expression we presented. If one is to point-split, infinitesimal segments of loop should be used to connect the split points to preserve gauge invariance and a more careful calculation would be in order.

What does all this tell us about the physical relevance of the solutions? The situation is remarkably similar to the one present in the loop representation of the free Maxwell field [11]. In that case, as here, there are two terms in the Hamiltonian that need to be regularized in a different way (in the case of gravity, the determinant of the metric requires splitting three points whereas the Hamiltonian only needs two). As a consequence of this, it is not surprising that the wave functions that solve the constraint have some regularization dependence. In the case of the Maxwell field the vacuum in the loop representation needs to be regularized. In fact, its form is exactly the same as that of the exponential of the Gauss linking number if one replaces the propagator of the Chern–Simons theory present in the latter by the propagator of the Maxwell field. This similarity is remarkable. The problem is therefore the same; the wave functions inherit regularization dependence since the regulator does not appear as an overall factor of the wave equation.

How could these regularization ambiguities be cured? In the Maxwell case they are solved by considering an 'extended' loop representation in which one allows the quantities X^{ax} to become smooth vector densities on the manifold without reference to any particular loop [12]. In the gravitational case such construction is being actively pursued [13], although it is more complicated. It is in this context that the present solutions really make sense. If one allows the X to become smooth functions the framing problem disappears and one is left with a solution that is a function of vector fields and only reduces to the Gauss linking number in a very special (singular) limit. It has been proved that the extensions of the Kauffman bracket and Jones polynomials to the case of smooth density fields are solutions of the extended constraints. A similar proof goes through for the extended Gauss linking number. In the extended representation, there are additional multivector densities needed in the representation. The 'Abelian' limit of the Kauffman bracket (the Gauss linking number) appears as the restriction of the 'extended' Kauffman bracket to the case in which higher-order multivector densities vanish. It would be interesting to study if such a limit could be pursued in a systematic way order by order. It would certainly provide new insights into how to construct non-perturbative quantum states of the gravitational field.

Acknowledgements

This paper is based on a talk given by J P at the Riverside conference. J P wishes to thank John Baez for inviting him to participate in the conference and hospitality in Riverside. This work was supported by grants NSF-PHY-92-07225, NSF-PHY93-96246, and by research funds of the University of Utah and Pennsylvania State University. Support from PEDECIBA (Uruguay) is also acknowledged. R G wishes to thank Karel Kuchař and Richard Price for hospitality at the University of Utah where part of this work was accomplished.

Bibliography

1. T. Jacobson, L. Smolin, Nucl. Phys. **B299**, 295 (1988).

2. C. Rovelli, L. Smolin, Phys. Rev. Lett. **61**, 1155 (1988); Nucl. Phys. **B331**, 80 (1990).

3. B. Brügmann, J. Pullin, Nucl. Phys. **B363**, 221 (1991).

4. R. Gambini, Phys. Lett. **B255**, 180 (1991).

5. B. Brügmann, R. Gambini, J. Pullin, Phys. Rev. Lett. **68**, 431 (1992).

6. E. Witten, Commun. Math. Phys. **121**, 351 (1989).

7. E. Guadagnini, M. Martellini, M. Mintchev, Nucl. Phys. **B330**, 575 (1990).

8. B. Brügmann, R. Gambini, J. Pullin, General Relativity Gravitation **25**, 1 (1993).

9. J. Griego, "The extended loop representation: a first approach", preprint (1993).

10. R. Gambini, A. Trias, Phys. Rev. **D27**, 2935 (1983).

11. C. Di Bartolo, F. Nori, R. Gambini, L. Leal, A. Trias, Lett. Nuovo. Cim. **38**, 497 (1983).

12. D. Armand-Ugon, R. Gambini, J. Griego, L. Setaro, "Classical loop actions of gauge theories", preprint hep-th:9307179 (1993).

13. C. Di Bartolo, R. Gambini, J. Griego, J. Pullin, in preparation.

Vassiliev Invariants and the Loop States in Quantum Gravity

Louis H. Kauffman

Department of Mathematics, Statistics, and Computer Science,
University of Illinois at Chicago,
Chicago, Illinois 60607, USA
(email: U10451@uicvm.uic.edu)

Abstract

This paper derives (at the physical level of rigor) a general switching formula for the link invariants defined via the Witten functional integral. The formula is then used to evaluate the top row of the corresponding Vassiliev invariants and it is shown to match the results of Bar-Natan obtained via perturbation expansion of the integral. It is explained how this formalism for Vassiliev invariants can be used in the context of the loop transform for quantum gravity.

1 Introduction

The purpose of this paper is to expose properties of Vassiliev invariants by using the simplest of the approaches to the functional integral definition of link invariants. These methods are strong enough to give the top row evaluations of Vassiliev invariants for the classical Lie algebras. They give an insight into the structure of these invariants without using the full perturbation expansion of the integral. One reason for examining the invariants in this light is the possible applications to the loop variables approach to quantum gravity. The same level of handling the functional integral is commonly used in the loop transform for quantum gravity.

The paper is organized as follows. Section 2 is a brief discussion of the nature of the functional integral. Section 3 details the formalism of the functional integral that we shall use, and works out a difference formula for changing crossings in the link invariant and a formula for the change of framing. These are applied to the case of an $SU(N)$ gauge group. Section 4 defines the Vassiliev invariants and shows how to formulate them in terms of the functional integral. In particular we derive the specific expressions for the 'top rows' of Vassiliev invariants corresponding to the fundamental representation of $SU(N)$. This gives a neat point of view on the results of Bar-Natan, and also gives a picture of the structure of the graphical vertex associated with the Vassiliev invariant. We see that this vertex is not just a transversal intersection of Wilson loops, but rather has the structure

77

of Casimir insertion (up to first order of approximation) coming from the
difference formula in the functional integral. This clarifies an issue raised
by John Baez in [4]. Section 5 is a quick remark about the loop formalism
for quantum gravity and its relationships with the invariants studied in the
previous sections. This marks the beginning of a study that will be carried
out in detail elsewhere.

2   Quantum mechanics and topology

In [47] Edward Witten proposed a formulation of a class of 3-manifold
invariants as generalized Feynman integrals taking the form $Z(M)$ where

$$Z(M) = \int dA \exp\left[(ik/4\pi)S(M, A)\right].$$

Here M denotes a 3-manifold without boundary and A is a gauge field (also
called a gauge potential or gauge connection) defined on M. The gauge
field is a 1-form on M with values in a representation of a Lie algebra.
The group corresponding to this Lie algebra is said to be the *gauge group*
for this particular field. In this integral the 'action' $S(M, A)$ is taken to
be the integral over M of the trace of the Chern-Simons three-form $CS = AdA + (2/3)AAA$. (The product is the wedge product of differential forms.)

Instead of integrating over paths, the integral $Z(M)$ integrates over
all gauge fields modulo gauge equivalence (see [2] for a discussion of the
definition and meaning of gauge equivalence). This generalization from
paths to fields is characteristic of quantum field theory.

Quantum field theory was designed in order to accomplish the quantization of electromagnetism. In quantum electrodynamics the classical
entity is the electromagnetic field. The question posed in this domain is to
find the value of an amplitude for starting with one field configuration and
ending with another. The analogue of all paths from point a to point b is
'all fields from field A to field B'.

Witten's integral $Z(M)$ is, in its form, a typical integral in quantum
field theory. In its content $Z(M)$ is highly unusual. The formalism of
the integral, and its internal logic, support the existence of a large class
of topological invariants of 3-manifolds and associated invariants of knots
and links in these manifolds.

The invariants associated with this integral have been given rigorous
combinatorial descriptions (see [35, 43, 27, 29, 25]), but questions and conjectures arising from the integral formulation are still outstanding (see [13]).

3   Links and the Wilson loop

We now look at the formalism of the Witten integral and see how it im-
plicates invariants of knots and links corresponding to each classical Lie
algebra. We need the *Wilson loop*. The Wilson loop is an exponentiated

version of integrating the gauge field along a loop K in 3-space that we take to be an embedding (knot) or a curve with transversal self-intersections. For the purpose of this discussion, the Wilson loop will be denoted by the notation $\langle K|A \rangle$ to denote the dependence on the loop K and the field A. It is usually indicated by the symbolism $\mathrm{Tr}(P \exp(\oint KA))$.

Thus $\langle K|A \rangle = \mathrm{Tr}(P \exp(\oint KA))$. Here the P denotes path-ordered integration. The symbol Tr denotes the trace of the resulting matrix.

With the help of the Wilson loop functional on knots and links, Witten [47] writes down a functional integral for link invariants in a 3-manifold M:

$$Z(M, K) = \int dA \exp[(ik/4\pi)S(M, A)] \mathrm{Tr}\left(P \exp\left(\oint KA\right)\right)$$

$$= \int dA \exp[(ik/4\pi)S]\langle K|A \rangle.$$

Here $S(M, A)$ is the Chern–Simons action, as in the previous discussion. We abbreviate $S(M, A)$ as S. Unless otherwise mentioned, the manifold M will be the 3-dimensional sphere S^3.

An analysis of the formalism of this functional integral reveals quite a bit about its role in knot theory. This analysis depends upon key facts relating the curvature of the gauge field to both the Wilson loop and the Chern–Simons Lagrangian. To this end, let us recall the local coordinate structure of the gauge field $A(x)$, where x is a point in 3-space. We can write $A(x) = A_k^a(x)T_a dx^k$ where the index a ranges from 1 to m with the Lie algebra basis $\{T_1, T_2, T_3, \ldots, T_m\}$. The index k goes from 1 to 3. For each choice of a and k, $A_a^k(x)$ is a smooth function defined on 3-space. In $A(x)$ we sum over the values of repeated indices. The Lie algebra generators T_a are actually matrices corresponding to a given representation of an abstract Lie algebra. We assume some properties of these matrices as follows:

1. $[T_a, T_b] = if_{abc}T_c$ where $[x, y] = xy - yx$, and f_{abc} (the matrix of structure constants) is totally antisymmetric. There is summation over repeated indices.

2. $\mathrm{Tr}(T_a T_b) = \delta_{ab}/2$ where δ_{ab} is the Kronecker delta ($\delta_{ab} = 1$ if $a = b$ and zero otherwise).

We also assume some facts about curvature. (The reader may compare with the exposition in [24]. But note the difference in conventions on the use of i in the Wilson loops and curvature definitions.) The first fact is the relation of Wilson loops and curvature for small loops:

Fact 1 The result of evaluating a Wilson loop about a very small planar circle around a point x is proportional to the area enclosed by this circle times the corresponding value of the curvature tensor of the gauge field evaluated at x. The curvature tensor is written $F_{rs}^a(x)T_a dx^r dy^s$. It is the local coordinate expression of $dA + AA$.

Application of Fact 1 Consider a given Wilson loop $\langle K|A\rangle$. Ask how its value will change if it is deformed infinitesimally in the neighborhood of a point x on the loop. Approximate the change according to Fact 1, and regard the point x as the place of curvature evaluation. Let $\delta\langle K|A\rangle$ denote the change in the value of the loop. $\delta\langle K|A\rangle$ is given by the formula $\delta\langle K|A\rangle = dx^r dx^s F^a_{rs}(x)T_a\langle K|A\rangle$. This is the first-order approximation to the change in the Wilson loop.

In this formula it is understood that the Lie algebra matrices T_a are to be inserted into the Wilson loop at the point x, and that we are summing over repeated indices. This means that each $T_a\langle K|A\rangle$ is a new Wilson loop obtained from the original loop $\langle K|A\rangle$ by leaving the form of the loop unchanged, but inserting the matrix T_a into the loop at the point x. See the figure below.

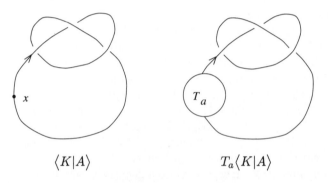

$$\langle K|A\rangle \qquad\qquad\qquad T_a\langle K|A\rangle$$

Remark on insertion The Wilson loop is the limit, over partitions of the loop K, of products of the matrices $\left(1 + A(x)\right)$ where x runs over the partition. Thus one can write symbolically,

$$\langle K|A\rangle = \prod_{x\in K}\left(1 + A(x)\right) = \prod_{x\in K}\left(1 + A^a_k(x)T_a dx^k\right).$$

It is understood that a product of matrices around a closed loop connotes the trace of the product. The ordering is forced by the 1-dimensional nature of the loop. Insertion of a given matrix into this product at a point on the loop is then a well-defined concept. If T is a given matrix then *it is understood that $T\langle K|A\rangle$ denotes the insertion of T into some point of the loop.* In the case above, it is understood from the context of the formula $ds^r dx^s F^a_{rs}(x)T_a\langle K|A\rangle$ that the insertion is to be performed at the point x indicated in the argument of the curvature.

Remark The previous remark implies the following formula for the variation of the Wilson loop with respect to the gauge field:

$$\frac{\delta\langle K|A\rangle}{\delta\left(A_k^a(x)\right)} = dx^k T_a \langle K|A\rangle.$$

Varying the Wilson loop with respect to the gauge field results in the insertion of an infinitesimal Lie algebra element into the loop.

Proof

$$\begin{aligned}
\frac{\delta\langle K|A\rangle}{\delta\left(A_k^a(x)\right)} &= \frac{\delta}{\delta\left(A_k^a(x)\right)} \prod_{y\in K}\left(1 + A_k^a(y)T_a dy^k\right) \\
&= \left[\prod_{y<x}\left(1 + A_k^a(y)T_a dy^k\right)\right][T_a dx^k]\left[\prod_{y>x}\left(1 + A_k^a(y)T_a dy^k\right)\right] \\
&= dx^k T_a \langle K|A\rangle.
\end{aligned}$$

$\square$

Fact 2 The variation of the Chern–Simons action S with respect to the gauge potential at a given point in 3-space is related to the values of the curvature tensor at that point by the following formula (where ϵ_{abc} is the epsilon symbol for three indices):

$$F_{rs}^a(x) = \epsilon_{rst}\frac{\delta S}{\delta\left(A_t^a(x)\right)}.$$

With these facts at hand we are prepared to determine how the Witten integral behaves under a small deformation of the loop K.

Proposition 1 (Compare [24].) *All statements of equality in this proposition are up to order* $(1/k)^2$.

1. Let $Z(K) = Z(K, S^3)$ and let $\delta Z(K)$ denote the change of $Z(K)$ under an infinitesimal change in the loop K. Then

$$\delta Z(K) = (4\pi i/k)\int dA \exp\left[(ik/4\pi)S\right][\epsilon_{rst}dx^r dy^s dz^t]T_a T_a\langle K|A\rangle.$$

The sum is taken over repeated indices, and the insertion is taken of the matrix products $T_a T_a$ at the chosen point x on the loop K that is regarded as the 'center' of the deformation. The volume element $[\epsilon_{rst}dx^r dy^s dz^t]$ is taken with regard to the infinitesimal directions of the loop deformation from this point on the original loop.

2. The same formula applies, with a different interpretation, to the case where x is a double point of transversal self-intersection of a loop

K, and the deformation consists in shifting one of the crossing segments perpendicularly to the plane of intersection so that the self-intersection point disappears. In this case, one T_a is inserted into each of the transversal crossing segments so that $T_a T_a \langle K|A \rangle$ denotes a Wilson loop with a self-intersection at x and insertions of T_a at $x + \epsilon_1$ and $x + \epsilon_2$, where ϵ_1 and ϵ_2 denote small displacements along the two arcs of K that intersect at x. In this case, the volume form is nonzero, with two directions coming from the plane of movement of one arc, and the perpendicular direction is the direction of the other arc.

Proof

$$
\begin{aligned}
Z(K) &= \int dA \exp\big[(ik/4\pi)S\big]\delta\langle K|A\rangle \\
&= \int dA \exp\big[(ik/4\pi)S\big]dx^r dy^s F^a_{rs}(x)T_a\langle K|A\rangle \qquad \text{(Fact 1)} \\
&= \int dA \exp\big[(ik/4\pi)S\big]dx^r dy^s \epsilon_{rst}\frac{\delta S}{\delta\big(A^a_t(x)\big)}T_a\langle K|A\rangle \qquad \text{(Fact 2)} \\
&= \int dA \exp\big[(ik/4\pi)S\big]\Big(\frac{\delta S}{\delta\big(A^a_t(x)\big)}\Big)\epsilon_{rst}dx^r dy^s T_a\langle K|A\rangle \\
&= (-4\pi i/k)\int dA\frac{\delta\big\{\exp\big[(ik/4\pi)S\big]\big\}}{\delta\big(A^a_t(x)\big)}\epsilon_{rst}dx^r dy^s T_a\langle K|A\rangle \\
&= (4\pi i/k)\int dA \exp\big[(ik/4\pi)S\big]\epsilon_{rst}dx^r dy^s \frac{\delta\big\{T_a\langle K|A\rangle\big\}}{\delta\big(A^a_t(x)\big)} \\
&\qquad\text{(integration by parts)} \\
&= (4\pi i/k)\int dA \exp\big[(ik/4\pi)S\big][\epsilon_{rst}dx^r dy^s dz^t]T_a T_a\langle K|A\rangle \\
&\qquad\text{(differentiating the Wilson loop).}
\end{aligned}
$$

This completes the formalism of the proof. In the case of part 2, the change of interpretation occurs at the point in the argument when the Wilson loop is differentiated. Differentiating a self-intersecting Wilson loop at a point of self-intersection is equivalent to differentiating the corresponding product of matrices at a variable that occurs at two points in the product (corresponding to the two places where the loop passes through the point). One of these derivatives gives rise to a term with volume form equal to zero, the other term is the one that is described in part 2. This completes the proof of the proposition.   □

Applying Proposition 1 As the formula of Proposition 1 shows, the integral $Z(K)$ is unchanged if the movement of the loop does not involve three independent space directions (since $\epsilon_{rst}dx^r dy^s dz^t$ computes a volume). This means that $Z(K) = Z(S^3, K)$ is invariant under moves that slide the knot along a plane. In particular, this means that if the knot K is given in the nearly planar representation of a knot diagram, then $Z(K)$ is invariant under regular isotopy of this diagram. That is, it is invariant under the Reidemeister moves II and III. We expect more complicated behavior under move I since this deformation does involve three spatial directions. This will be discussed momentarily.

We first determine the difference between $Z(K_+)$ and $Z(K_-)$ where K_+ and K_- denote the knots that differ only by switching a single crossing. We take the given crossing in K_+ to be the positive type, and the crossing in K_- to be of negative type.

$$K_+ \qquad\qquad K_- \qquad\qquad K_\#$$

The strategy for computing this difference is to use $K_\#$ as an intermediate, where $K_\#$ *is the link with a transversal self-crossing replacing the given crossing in* K_+ *or* K_-. Thus we must consider $\Delta_+ = Z(K_+) - Z(K_\#)$ and $\Delta_- = Z(K_-) - Z(K_\#)$. The second part of Proposition 1 applies to each of these differences and gives

$$\Delta_+ = (4\pi i/k) \int dA \exp\left[(ik/4\pi)S\right] [\epsilon_{rst}dx^r dy^s dz^t] T_a T_a \langle K_\# | A \rangle$$

where, by the description in Proposition 1, this evaluation is taken along the loop $K_\#$ with the singularity and the $T_a T_a$ insertion occurs along the two transversal arcs at the singular point. The sign of the volume element will be opposite for Δ_- and consequently we have that

$$\Delta_+ + \Delta_- = 0.$$

(The volume element $[\epsilon_{rst}dx^r dy^s dz^t]$ must be given a conventional value in our calculations. There is no reason to assign it different absolute values for the cases of Δ_+ and Δ_- since they are symmetric except for the sign.)

Therefore $Z(K_+) - Z(K_\#) + Z(K_-) - Z(K_\#) = 0$. Hence

$$Z(K_\#) = (1/2)\left[Z(K_+) + Z(K_-)\right].$$

This result is central to our further calculations. It tells us that the evaluation of a singular Wilson loop can be replaced with the average of the results of resolving the singularity in the two possible ways.

Now we are interested in the difference $Z(K_+) - Z(K_-)$:

$$Z(K_+) - Z(K_-) = \Delta_+ - \Delta_- = 2\Delta_+$$

$$= (8\pi i/k) \int dA \exp\left[(ik/4\pi)S\right]\left[\epsilon_{rst}dx^r dy^s dz^t\right]T_aT_a\langle K_\#|A\rangle.$$

Volume convention It is useful to make a specific convention about the volume form. We take

$$\left[\epsilon_{rst}dx^r dy^s dz^t\right] = \frac{1}{2} \text{ for } \Delta_+ \text{ and } -\frac{1}{2} \text{ for } \Delta_-.$$

Thus

$$Z(K_+) - Z(K_-) = (4\pi i/k) \int dA \exp\left[(ik/4\pi)S\right]T_aT_a\langle K_\#|A\rangle.$$

Integral notation Let $Z(T_aT_aK_\#)$ denote the integral

$$Z(T_aT_aK_\#) = \int dA \exp\left[(ik/4\pi)S\right]T_aT_a\langle K_\#|A\rangle.$$

Difference formula Write the difference formula in the abbreviated form

$$Z(K_+) - Z(K_-) = (4\pi i/k)Z(T_aT_aK_\#).$$

This formula is the key to unwrapping many properties of the knot invariants. For diagrammatic work it is convenient to rewrite the difference equation in the form shown below. The crossings denote small parts of otherwise identical larger diagrams, and the Casimir insertion $T_aT_aK_\#$ is indicated with crossed lines entering a disk labelled C.

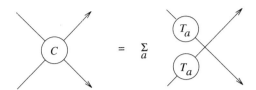

The Casimir The element $\sum_a T_a T_a$ of the universal enveloping algebra is called the Casimir. Its key property is that it is in the center of the algebra. Note that by our conventions $\text{Tr}(\sum_a T_a T_a) = \sum_a \Delta_{aa}/2 = d/2$ where d *is the dimension of the Lie algebra.* This implies that an unknotted loop with one singularity and a Casimir insertion will have Z-value $d/2$.

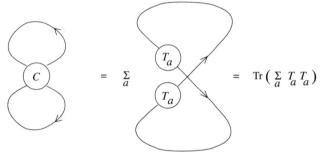

In fact, for the classical semisimple Lie algebras *one can choose a basis so that the Casimir is a diagonal matrix with identical values* $(d/2D)$ *on its diagonal.* D is the dimension of the representation. We then have the general formula $Z(T_a T_a K_\#^{\text{loc}}) = (d/2D)Z(K)$ for any knot K. Here $K_\#^{\text{loc}}$ denotes the singular knot obtained by placing a local self-crossing loop in K as shown below:

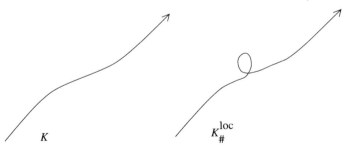

Note that $Z(K_\#^{\text{loc}}) = Z(K)$. (Let the flat loop shrink to nothing. The Wilson loop is still defined on a loop with an isolated cusp and it is equal to the Wilson loop obtained by smoothing that cusp.)

Let K_+^{loc} denote the result of adding a positive local curl to the knot K, and K_-^{loc} the result of adding a negative local curl to K.

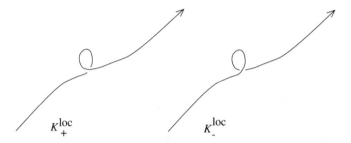

$$K_+^{\mathrm{loc}} \qquad\qquad K_-^{\mathrm{loc}}$$

Then by the above discussion and the difference formula, we have

$$
\begin{aligned}
Z(K_+^{\mathrm{loc}}) &= Z(K_\#^{\mathrm{loc}}) + (2\pi i/k)Z(T_a T_a K_\#^{\mathrm{loc}}) \\
&= Z(K) + (2\pi i/k)(d/2D)Z(K).
\end{aligned}
$$

Thus,

$$
Z(K_+^{\mathrm{loc}}) = \left[1 + (\pi i/k)(d/D)\right]Z(K).
$$

Similarly,

$$
Z(K_-^{\mathrm{loc}}) = \left[1 - (\pi i/k)(d/D)\right]Z(K).
$$

These calculations show how the difference equation, the Casimir, and properties of Wilson loops determine the framing factors for the knot invariants. In some cases we can use special properties of the Casimir to obtain skein relations for the knot invariant.

For example, in the fundamental representation of the Lie algebra for $SU(N)$ the Casimir obeys the following equation (see [24, 5]):

$$
\sum_a (T_a)_{ij}(T_a)_{kl} = \left(\frac{1}{2}\right)\delta_{il}\delta_{jk} - \left(\frac{1}{2N}\right)\delta_{ij}\delta_{kl}.
$$

Hence

$$
Z(T_a T_a K_\#) = \left(\frac{1}{2}\right)Z(K_0) - \left(\frac{1}{2N}\right)Z(K_\#)
$$

where K_0 denotes the result of smoothing a crossing as shown below:

$$K_0$$

Using $Z(K_\#) = [Z(K_+) + Z(K_-)]/2$ and the difference identity, we obtain

$$Z(K_+) - Z(K_-) = (4\pi i/k)\left(Z(K_0)/2 - \left(\frac{1}{2N}\right)\{[Z(K_+) + Z(K_-)]/2\}\right).$$

Hence

$$(1 + \pi i/Nk)Z(K_+) - (1 - \pi i/Nk)Z(K_-) = (2\pi i/k)Z(K_0)$$

or

$$e\left(\frac{1}{N}\right)Z(K_+) - e\left(-\frac{1}{N}\right)Z(K_-) = [e(1) - e(-1)]Z(K_0)$$

where $e(x) = \exp[(\pi i/k)x]$ taken up to $O(1/k^2)$.

Here $d = N^2 - 1$ and $D = N$, so the framing factor is

$$\alpha = \{1 + (\pi i/k)[(N^2 - 1)/N]\} = e[(N - (1/N)].$$

Therefore, if $P(K) = \alpha^{-w(K)} Z(K)$ denotes the normalized invariant of ambient isotopy associated with $Z(K)$ (with $w(K)$ the sum of the crossing signs of K), then

$$\alpha e(1/N)P(K_+) - \alpha^{-1} e(-1/N)P(K_-) = [e(1) - e(-1)]P(K_0).$$

Hence

$$e(N)P(K_+) - e(-N)P(K_-) = [e(1) - e(-1)]P(K_0).$$

This last equation shows that $P(K)$ is a specialization of the Homfly polynomial for arbitrary N, and that for $N = 2$ $(SU(2))$ it is a specialization of the Jones polynomial.

4 Graph invariants and Vassiliev invariants

We now apply this integral formalism to the structure of rigid vertex graph invariants that arise naturally in the context of knot polynomials. If $V(K)$ is a (Laurent-polynomial-valued, or, more generally, commutative-ring-valued) invariant of knots, then it can be naturally extended to an invariant of rigid vertex graphs by defining the invariant of graphs in terms of the knot invariant via an 'unfolding' of the vertex as indicated below [26]:

$$V(K_\$) = aV(K_+) + bV(K_-) + cV(K_0).$$

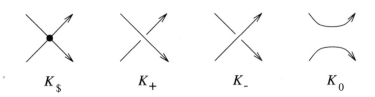

Here $K_\$$ indicates an embedding with a transversal 4-valent vertex ($). We use the symbol \$ to distinguish this choice of vertex designation from the previous one involving a self-crossing Wilson loop.

Formally, this means that we define $V(G)$ for an embedded 4-valent graph G by taking the sum over $a^{i+(S)}b^{i-(S)}c^{io(S)}V(S)$ for all knots S obtained from G by replacing a node of G either with a crossing of positive or negative type, or with a smoothing (denoted 0). It is not hard to see that if $V(K)$ is an ambient isotopy invariant of knots, then this extension is a rigid vertex isotopy invariant of graphs. In rigid vertex isotopy the cyclic order at the vertex is preserved, so that the vertex behaves like a rigid disk with flexible strings attached to it at specific points.

There is a rich class of graph invariants that can be studied in this manner. The *Vassiliev invariants* [5, 6, 44] constitute the important special case of these graph invariants where $a = +1$, $b = -1$, and $c = 0$. Thus $V(G)$ is a Vassiliev invariant if

$$V(K_\$) = V(K_+) - V(K_-).$$

$V(G)$ is said to be the *finite type* k *if* $V(G) = 0$ *whenever* $\#(G) > k$ where $\#(G)$ denotes the number of 4-valent nodes in the graph G. If V is the finite type k, then $V(G)$ *is independent of the embedding type of the graph* G *when* G *has exactly* k *nodes*. This follows at once from the definition of finite type. The values of $V(G)$ on all the graphs of k nodes is called the *top row* of the invariant V.

For purposes of enumeration it is convenient to use chord diagrams to enumerate and indicate the abstract graphs. A chord diagram consists of an oriented circle with an even number of points marked along it. These points are paired with the pairing indicated by arcs or chords connecting the paired points. See the figure below.

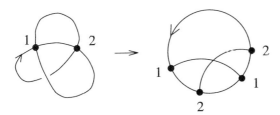

This figure illustrates the process of associating a chord diagram to a given embedded 4-valent graph. Each transversal self-intersection in the embedding is matched to a pair of points in the chord diagram.

5 Vassiliev invariants from the functional integral

In order to examine the Vassiliev invariants associated with the functional integral, we must first normalize these invariants to invariants of ambient isotopy, and then consider the structure of the difference between positive and negative crossings. The framed difference formula provides the necessary information for obtaining the top row.

We have shown that $Z(K_+^{\text{loc}}) = \alpha Z(K)$ with $\alpha = e(d/D)$. Hence

$$P(K) = \alpha^{-w(K)} Z(K)$$

is an ambient isotopy invariant. The equation

$$Z(K_+) - Z(K_-) = (4\pi i/k) Z(T_a T_a K_\#)$$

implies that if $w(K_+) = w + 1$, then we have the *ambient isotopy difference formula:*

$$P(K_+) - P(K_-) = \alpha^{-w}(4\pi i/k)\big[Z(T_a T_a K_\#) - (d/2D)Z(K_\#)\big].$$

We leave the proof of this formula as an exercise for the reader.

This formula tells us that for the Vassiliev invariant associated with P we have

$$P(K_\$) = \alpha^{-w}(4\pi i/k)\big[Z(T_a T_a K_\#) - (d/2D)Z(K_\#)\big].$$

Furthermore, if $V_j(K)$ denotes the coefficient of $(4\pi i/k)^j$ in the expansion of $P(K)$ in powers of $(1/k)$, then the ambient difference formula implies that $(1/k)^j$ divides $P(G)$ when G has j or more nodes. Hence $V_j(G) = 0$ if G has more than j nodes. Therefore $V_j(K)$ is a Vassiliev invariant of finite type. (This result was proved by Birman and Lin [6] by different methods and by Bar-Natan [5] by methods equivalent to ours.)

The fascinating thing is that the ambient difference formula, appropriately interpreted, actually tells us how to compute $V_k(G)$ when G has k

nodes. Under these circumstances each node undergoes a Casimir insertion, and because the Wilson loop is being evaluated abstractly, independent of the embedding, we insert nothing else into the loop. Thus we take the pairing structure associated with the graph (the so-called chord diagram) and use it as a prescription for obtaining a trace of a sum of products of Lie algebra elements with T_a and T_a inserted for each pair or a simple crossover for the pair multiplied by $(d/2D)$. This yields the graphical evaluation implied by the recursion

$$V(G_\$) = \left[V(T_aT_aG_\#) - (d/2D)V(G_\#)\right].$$

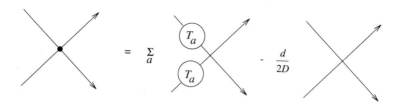

At each stage in the process one node of G disappears or it is replaced by these insertions. After k steps we have a fully inserted sum of abstract Wilson loops, each of which can be evaluated by taking the indicated trace. This result is equivalent to Bar-Natan's result, but it is very interesting to see how it follows from a minimal approach to the Witten integral.

In particular, it follows from Bar-Natan [5] and Kontsevich [28] that the condition of topological invariance is translated into the fact that the Lie bracket is represented as a commutator *and* that it is closed with respect to the Lie algebra. Diagrammatically we have:

Since

$$T_aT_b - T_bT_a = \sum_c if_{abc}T_c$$

we obtain

This relationship on chord diagrams is the seed of all the topology. In particular, it implies the basic 4-term relation

Proof

The presence of this relation on chord diagrams for $V_i(G)$ with $\#(G) = i$ is the basis for the existence of a corresponding Vassiliev invariant. There is not room here to go into more detail about this matter, and so we bring this discussion to a close. Nevertheless, it must be mentioned that this brings us to the core of the main question about Vassiliev invariants: are there non-trivial Vassiliev invariants of knots and links that cannot be constructed through combinations of Lie algebraic invariants? There are many other open questions in this arena, all circling this basic problem.

6   Quantum gravity—loop states

We now discuss the relationship of Wilson loops and quantum gravity that is forged in the theory of Ashtekar, Rovelli, and Smolin [1]. In this theory the metric is expressed in terms of a spin connection A, and quantization involves considering wave functions $\psi(A)$. Smolin and Rovelli analyze the loop transform $\hat{\psi}(K) = \int dA\psi(A)\langle K|A\rangle$ where $\langle K|A\rangle$ denotes the Wilson loop for the knot or singular embedding K. Differential operators on the wave function can be referred, via integration by parts, to corresponding statements about the Wilson loop. It turns out that the condition that $\hat{\psi}(K)$ be a knot invariant (without framing dependence) is equivalent to the so-called diffeomorphism constraint [37] for these wave functions. In this way, knots and weaves and their topological invariants become a language for representing a state of quantum gravity.

The main point that we wish to make in the relationship of the Vassiliev invariants to these loop states is that the Vassiliev vertex

satisfying

$$\times\!\!\!\bullet = \times\!\!\!\times - \times\!\!\!\times$$

is not simply a transverse intersection of Wilson loops. We have seen that it follows from the difference formula that the Vassiliev vertex is much more complex than this—that it involves the Casimir insertion at the transverse intersection up to the first order of approximation, and that this structure can be used to compute the top row of the corresponding Vassiliev invariant. This situation suggests that one should amalgamate the formalism of the Vassiliev invariants with the structure of the Poisson algebras of loops and insertions used in the quantum gravity theory [37, 33]. It also suggests taking the formalism of these invariants (in the functional integral form) quite seriously even in the absence of an appropriate measure theory.

One can begin to work backwards, taking the position that invariants that do not ostensibly satisfy the diffeomorphism constraint (owing to change of value under framing change) nevertheless still define states of a quantum gravity theory that is a modification of the Ashtekar formulation. This theory can be investigated by working the transform methods backwards—from knots and links to differential operators and differential geometry.

All these remarks are the seeds for another paper. We close here, and ask the reader to stay tuned for further developments.

Acknowledgements

It gives the author pleasure to acknowledge the support of NSF Grant Number DMS 9205277 and the Program for Mathematics and Molecular Biology of the University of California at Berkeley, Berkeley, CA.

Bibliography

1. A. Ashtekar, C. Rovelli, and L. Smolin, Weaving a classical geometry with quantum threads. (Preprint 1992).

2. M. F. Atiyah, *Geometry of Yang–Mills Fields*, Accademia Nazionale dei Lincei Scuola Superiore Lezioni Fermiare, Pisa (1979).

3. M. F. Atiyah, *The Geometry and Physics of Knots*, Cambridge University Press (1990).

4. J. Baez, Link invariants of finite type and perturbation theory, *Lett. Math. Phys.* **26** (1991) 43–51.

5. D. Bar-Natan, On the Vassiliev knot invariants. (Preprint 1992).

6. J. Birman and X. S. Lin, Knot polynomials and Vassiliev's invariants, *Invent. Math.* to appear.

7. B. Brügmann, R. Gambini, and J. Pullin, Knot invariants as nondegenerate quantum geometries, *Phys. Rev. Lett.* **68** (1992) 431–434.

8. L. Crane, Conformal field theory, spin geometry and quantum gravity, *Phys. Lett.* **B259** (1991) 243–248.

9. L. Crane and D. Yetter, A categorical construction of 4D topological quantum field theories. (Preprint 1993).

10. P. A. M. Dirac, *Principles of Quantum Mechanics*, Oxford University Press (1958).

11. R. Feynman and A. R. Hibbs, *Quantum Mechanics and Path Integrals*, McGraw-Hill (1965).

12. D. Freed and R. Gompf, Computer calculations of Witten's 3-manifold invariants, *Commun. Math. Phys.* **41** (1991) 79–117.

13. S. Garoufalidis, Applications of TQFT to invariants in low dimensional topology. (Preprint 1993).

14. B. Hasslacher and M. J. Perry, Spin networks are simplicial quantum gravity, *Phys. Lett.* **B103** (1981) 21–24.

15. L. C. Jeffrey, On Some Aspects of Chern–Simons Gauge Theory. (Thesis, Oxford (1991)).

16. V. F. R. Jones, Index for subfactors, *Invent. Math.* **72** (1983) 1–25.

17. V. F. R. Jones, A polynomial invariant for links via von Neumann algebras, *Bull. Am. Math. Soc.* **129** (1985) 103–112.

18. V. F. R. Jones, A new knot polynomial and von Neumann algebras, *Not. Am. Math. Soc.* **33** (1986) 219–225.

19. V. F. R. Jones, Hecke algebra representations of braid groups and link polynomials, *Ann. Math.* **126** (1987) 335–338.

20. V. F. R. Jones, On knot invariants related to some statistical mechanics models, *Pacific J. Math.* **137** (1989) 311–334.

21. L. H. Kauffman, State models and the Jones polynomial, *Topology* **26** (1987) 395–407.

22. L. H. Kauffman, *On Knots*, Annals of Mathematics Studies Number 115, Princeton University Press (1987).

23. L. H. Kauffman, Statistical mechanics and the Jones polynomial, *AMS Contemp. Math. Ser.* **78** (1989) 263–297.

24. L. H. Kauffman, *Knots and Physics*, World Scientific Pub. (1991).

25. L. H. Kauffman and S. Lins, Temperley Lieb Recoupling Theory and Invariants of 3-Manifolds (to appear as Annals monograph, Princeton University Press).

26. L. H. Kauffman and P. Vogel, Link polynomials and a graphical calculus, *J. Knot Theor. Ramifications,* **1** (1992) 59–104.

27. R. Kirby and P. Melvin, On the 3-manifold invariants of Reshetikhin-Turaev for sl$(2, C)$, *Invent. Math.* **105** (1991) 473–545.

28. M. Kontsevich, Graphs, homotopical algebra and low dimensional topology. (Preprint 1992).

29. W. B. R. Lickorish, 3-manifolds and the Temperley Lieb algebra, *Math. Ann.* **290** (1991) 657–670.

30. C. W. Misner, K. S. Thorne, and J. A. Wheeler, *Gravitation*, W. H. Freeman (1973).

31. H. Ooguri, Discrete and continuum approaches to three-dimensional quantum gravity. (Preprint 1991).

32. H. Ooguri, Topological lattice models in four dimensions, *Mod. Phys. Lett.* **A7** (1992) 2799–2810.

33. J. Pullin, Knot theory and quantum gravity – a primer. (Preprint

1993).

34. N. Y. Reshetikhin and V. Turaev, Ribbon graphs and their invariants derived from quantum groups, *Commun. Math. Phys.* **127** (1990) 1–26.

35. N. Y. Reshetikhin and V. Turaev, Invariants of three manifolds via link polynomials and quantum groups, *Invent. Math.* **103** (1991) 547–597.

36. L. Rozansky, Large k asymptotics of Witten's invariant of Seifert manifolds. (Preprint 1993).

37. L. Smolin, Quantum gravity in the self-dual representation, *Contemp. Math.* **71** (1988) 55–97.

38. T. Stanford, Finite-type invariants of knots, links and graphs. (Preprint 1992).

39. V. G. Turaev, Quantum invariants of links and 3-valent graphs in 3-manifolds. (Preprint 1990).

40. V. G. Turaev, Quantum invariants of 3-manifolds and a glimpse of shadow topology. (Preprint 1990).

41. V. G. Turaev, Topology of shadows. (Preprint 1992).

42. V. G. Turaev and O. Viro, State sum invariants of 3-manifolds and quantum $6j$ symbols, *Topology* **31** (1992) 865–902.

43. V. G. Turaev and H. Wenzl, Quantum invariants of 3-manifolds associated with classical simple Lie algebras. (Preprint).

44. V. Vassiliev, Cohomology of knot spaces. In: *Theory of Singularities and its Applications* (V. I. Arnold, ed.), Am. Math. Soc. (1990), pp. 23–69.

45. K. Walker, On Witten's 3-manifold invariants. (Preprint 1991).

46. R. Williams and F. Archer, The Turaev–Viro state sum model and 3-dimensional quantum gravity. (Preprint 1991).

47. E. Witten, Quantum field theory and the Jones polynomial, *Commun. Math. Phys.* **121** (1989) 351–399.

48. E. Witten, 2+1 dimensional gravity as an exactly soluble system, *Nucl. Phys.* **B311** (1989), 46–78.

Geometric Structures and Loop Variables in (2+1)-dimensional Gravity

Steven Carlip

*Department of Physics, University of California,
Davis, California 95616, USA
(email: carlip@dirac.ucdavis.edu)*

Abstract

In this chapter, I review the relationship between the metric formulation of (2+1)-dimensional gravity and the loop observables of Rovelli and Smolin. I emphasize the possibility of reconstructing the classical geometry, via the theory of geometric structures, from the values of the loop variables. I close with a brief discussion of implications for quantization, particularly for covariant canonical approaches to quantum gravity.

1   Introduction

Ashtekar's connection representation for general relativity [1, 2] and the closely related loop variable approach [3] have generated a good deal of excitement over the past few years. While it is too early to make firm predictions, there seems to be some real hope that these new variables will allow the construction of a consistent non-perturbative quantum theory of gravity. Some important progress has been made: a number of observables have been found, large classes of quantum states have been identified, and the first steps have been taken towards establishing a reasonable weak-field perturbation theory [4].

Progress has been hampered, however, by the absence of a clear physical interpretation for the observables built out of Ashtekar's new variables. In part, the problem is simply one of unfamiliarity—physicists accustomed to metrics and their associated connections can find it difficult to make the transition to densitized triads and self-dual connections. But there is a deeper problem as well, inherent in almost any canonical formulation of general relativity. To define Ashtekar's variables, one must choose a time slicing, an arbitrary splitting of spacetime into spacelike hypersurfaces. But real geometry and physics cannot depend on such a choice; the true physical observables must somehow forget any details of the time slicing, and refer only to the invariant underlying geometry. To a certain extent, this is already a source of trouble in classical general relativity, where one must take care to separate physical phenomena from artifacts of coordinate choices. In canonical quantization, however, the problem becomes much sharper—all observables must be diffeomorphism invariants, and the need to reconstruct geometry and physics from such quantities becomes unavoidable.

In a sense, the Ashtekar program is a victim of its own success. For the first time, we can actually write down a large set of diffeomorphism-invariant operators, built out of the loop variables of Rovelli and Smolin in the 'loop representation' of quantum gravity [2, 3]. But although some progress has been made in defining area and volume operators in terms of these variables [5], the goal of reconstructing spacetime geometry from such invariant quantities remains out of reach.

The purpose of this chapter is to demonstrate that such a reconstruction is possible in the simple model of (2+1)-dimensional gravity, general relativity in two spatial dimensions plus time. A reduction in the number of dimensions greatly simplifies general relativity, allowing the use of powerful techniques not readily available in the realistic (3+1)-dimensional theory. As a consequence, many of the specific results presented here will not readily generalize to higher dimensions. But the success of (2+1)-dimensional gravity can be viewed as an 'existence proof' for canonical quantum gravity, and one may hope that at least some of the technical results have extensions to our physical spacetime.

2   From geometry to holonomies

Let us begin with a brief review of (2+1)-dimensional general relativity in first-order formalism. As our spacetime we take a 3-manifold M, which we shall often assume to have a topology $\mathbf{R} \times \Sigma$, where Σ is a closed orientable surface. The fundamental variables are a triad $e_\mu{}^a$—a section of the bundle of orthonormal frames—and a connection on the same bundle, which can be specified by a connection 1-form $\omega_\mu{}^a{}_b$.[10] The Einstein–Hilbert action can be written as

$$I_{\text{grav}} = \int_M e^a \wedge \left(d\omega_a + \frac{1}{2}\epsilon_{abc}\omega^b \wedge \omega^c \right), \qquad (2.1)$$

where $e^a = e_\mu{}^a dx^\mu$ and $\omega^a = \frac{1}{2}\epsilon^{abc}\omega_{\mu bc}dx^\mu$. The action is invariant under local $SO(2,1)$ transformations,

$$\delta e^a = \epsilon^{abc}e_b\tau_c \qquad (2.2)$$
$$\delta\omega^a = d\tau^a + \epsilon^{abc}\omega_b\tau_c,$$

---

[10]Indices μ, ν, ρ, ... are spacetime coordinate indices; i, j, k, ... are spatial coordinate indices; and a, b, c, ... are 'Lorentz indices', labelling vectors in an orthonormal basis. Lorentz indices are raised and lowered with the Minkowski metric η_{ab}. This notation is standard in papers in (2+1)-dimensional gravity, but differs from the usual conventions for Ashtekar variables, so readers should be careful in translation.

as well as 'local translations',

$$\delta e^a = d\rho^a + \epsilon^{abc}\omega_b\rho_c \qquad (2.3)$$
$$\delta\omega^a = 0.$$

I_{grav} is also invariant under diffeomorphisms of M, of course, but this is not an independent symmetry: Witten has shown [6] that when the triad $e_\mu{}^a$ is invertible, diffeomorphisms in the connected component of the identity are equivalent to transformations of the form (2.2)–(2.3). We therefore need only worry about equivalence classes of diffeomorphisms that are not isotopic to the identity, that is elements of the mapping class group of M.

The equations of motion coming from the action (2.1) are easily derived:

$$de^a + \epsilon^{abc}\omega_b \wedge e_c = 0 \qquad (2.4)$$

and

$$d\omega^a + \frac{1}{2}\epsilon^{abc}\omega_b \wedge \omega_c = 0. \qquad (2.5)$$

These equations have four useful interpretations, which will form the basis for our analysis:

1. We can solve (2.4) for ω as a function of e, and rewrite (2.5) as an equation for $\omega[e]$. The result is equivalent to the ordinary vacuum Einstein field equations,

$$R_{\mu\nu}[g] = 0, \qquad (2.6)$$

for the Lorentzian (i.e. pseudo-Riemannian) metric $g_{\mu\nu} = e_\mu{}^a e_\nu{}^b \eta_{ab}$. In 2+1 dimensions, these field equations are much more powerful than they are in 3+1 dimensions: the full Riemann curvature tensor is linearly dependent on the Ricci tensor,

$$R_{\mu\nu\rho\sigma} = g_{\mu\rho}R_{\nu\sigma} + g_{\nu\sigma}R_{\mu\rho} - g_{\nu\rho}R_{\mu\sigma} - g_{\mu\sigma}R_{\nu\rho} - \frac{1}{2}(g_{\mu\rho}g_{\nu\sigma} - g_{\mu\sigma}g_{\nu\rho})R, \qquad (2.7)$$

so (2.6) actually implies that the metric $g_{\mu\nu}$ is flat. The space of solutions of the field equations can thus be identified with the set of flat Lorentzian metrics on M.

2. As a second alternative, note that eqn (2.5) depends only on ω and not on e. In fact, (2.5) is simply the requirement that the curvature of ω vanish; that is, that ω be a flat $SO(2,1)$ connection. Moreover, eqn (2.4) can be interpreted as the statement that e is a cotangent vector to the space of flat $SO(2,1)$ connections; indeed, if $\omega(s)$ is a curve in the space of flat connections, the derivative of (2.5) gives

$$d\left(\frac{d\omega^a}{ds}\right) + \epsilon^{abc}\omega_b \wedge \left(\frac{d\omega_c}{ds}\right) = 0, \qquad (2.8)$$

which can be identified with (2.4) with

$$e^a = \frac{d\omega^a}{ds}.$$ (2.9)

To determine the physically inequivalent solutions of the field equations, we must still factor out the gauge transformations (2.2)–(2.3). The local Lorentz transformations (2.2) act on ω as ordinary $SO(2,1)$ gauge transformations, and tell us that only gauge equivalence classes of flat $SO(2,1)$ connections are relevant. Let us denote the space of such equivalence classes as $\tilde{\mathcal{N}}$. The transformations of e can once again be interpreted as statements about the cotangent space: if we consider a curve $\tau(s)$ of $SO(2,1)$ transformations, it is easy to check that the first equations of (2.2) and (2.3) follow from differentiating the second equation of (2.2), with

$$\rho^a = \frac{d\tau^a}{ds}$$ (2.10)

and e^a as in (2.9). A solution of the field equations is thus determined by a point in the cotangent bundle $T^*\tilde{\mathcal{N}}$.

Now, a flat connection on the frame bundle of M is determined by its holonomies, that is by a homomorphism

$$\rho : \pi_1(M) \to SO(2,1),$$ (2.11)

and gauge transformations act on ρ by conjugation. We can therefore write

$$\tilde{\mathcal{N}} = \text{Hom}(\pi_1(M), SO(2,1))/\sim,$$
$$\rho_1 \sim \rho_2 \text{ if } \rho_2 = h \cdot \rho_1 \cdot h^{-1}, \quad h \in SO(2,1).$$ (2.12)

It remains for us to factor out the diffeomorphisms that are not in the component of the identity, the mapping class group. These transformations act on $\tilde{\mathcal{N}}$ through their action as a group of automorphisms of $\pi_1(M)$, and in many interesting cases—for example, when M has the topology $\mathbf{R} \times \Sigma$—this action comprises the entire set of outer automorphisms of $\pi_1(M)$ [7]. If we denote equivalence under this action by $\sim'$, and let $\tilde{\mathcal{N}}/\sim' = \mathcal{N}$, we can express the space of solutions of the field equations (2.4)–(2.5) as[11] $T^*\mathcal{N}$.

When M has the topology $\mathbf{R} \times \Sigma$, this description can be further refined. In that case, the space $\tilde{\mathcal{N}}$—or at least the physically relevant connected component of $\tilde{\mathcal{N}}$—is homeomorphic to the Teichmüller

---

[11]Strictly speaking, one more subtlety remains. The space of homomorphisms (2.12) is not always connected, and it is often the case that only one connected component corresponds to physically admissible spacetimes. See [8, 9] for the mathematical structure and [6, 9, 10, 11] for physical implications.

space of Σ, and $\mathcal{N}$ is the corresponding moduli space [8, 12]. The set of vacuum spacetimes can thus be identified with the cotangent bundle of the moduli space of Σ, and many powerful results from Riemann surface theory become applicable.

3. A third approach is available if M has the topology $\mathbf{R} \times \Sigma$. For such a topology, it is useful to split the field equations into spatial and temporal components. Let us write

$$d = \tilde{d} + dt\, \partial_0,$$
$$e^a = \tilde{e}^a + e_0{}^a dt, \qquad (2.13)$$
$$\omega^a = \tilde{\omega}^a + \omega_0{}^a dt.$$

(This decomposition can be made in a less explicitly coordinate-dependent manner—see, for example, [13]—but the final results are unchanged.) The spatial projections of (2.4)–(2.5) take the same form as the original equations, with all quantities replaced by their 'tilded' spatial equivalents. As above, solutions may therefore be labelled by classes of homomorphisms, now from $\pi_1(\Sigma)$ to $SO(2,1)$, and the corresponding cotangent vectors. The temporal components of the field equations, on the other hand, now become

$$\partial_0\,\tilde{e}^a = \tilde{d}e_0{}^a + \epsilon^{abc}\tilde{\omega}_b e_{0c} + \epsilon^{abc}\tilde{e}_b \omega_{0c} \qquad (2.14)$$
$$\partial_0\,\tilde{\omega}^a = \tilde{d}\omega_0{}^a + \epsilon^{abc}\tilde{\omega}_b \omega_{0c}.$$

Comparing with (2.2)–(2.3), we see that the time development of $(\tilde{e}, \tilde{\omega})$ is entirely described by a gauge transformation, with $\tau^a = \omega_0{}^a$ and $\rho^a = e_0{}^a$.

This is consistent with our previous results, of course. For a topology of the form $\mathbf{R} \times \Sigma$, the fundamental group is simply that of Σ, and an invariant description in terms of holonomies should not be able to detect a particular choice of spacelike slice. Equation (2.14) shows in detail how this occurs: motion in coordinate time is merely a gauge transformation, and is therefore invisible to the holonomies. But the central dilemma described in the introduction now stands out sharply. For despite eqn (2.14), solutions of the (2+1)-dimensional field equations are certainly not static *as geometries*—they do not, in general, admit timelike Killing vectors. The real physical dynamics has somehow been hidden by this analysis, and must be uncovered if we are to find a sensible physical interpretation of our solutions. This puzzle is an example of the notorious 'problem of time' in gravity [14], and exemplifies one of the basic issues that must be resolved in order to construct a sensible quantum theory.

4. A final approach to the field equations (2.4)–(2.5) was suggested by Witten [6], who observed that the triad e and the connection ω could

be combined to form a single connection on an $ISO(2,1)$ bundle. $ISO(2,1)$, the 3-dimensional Poincaré group, has a Lie algebra with generators $\mathcal{J}^a$ and $\mathcal{P}^b$ and commutation relations

$$\left[\mathcal{J}^a, \mathcal{J}^b\right] = \epsilon^{abc}\mathcal{J}_c, \qquad \left[\mathcal{J}^a, \mathcal{P}^b\right] = \epsilon^{abc}\mathcal{P}_c, \qquad \left[\mathcal{P}^a, \mathcal{P}^b\right] = 0. \tag{2.15}$$

If we write a single connection 1-form

$$A = e^a \mathcal{P}_a + \omega^a \mathcal{J}_a \tag{2.16}$$

and define a 'trace,' an invariant inner product on the Lie algebra, by

$$\mathrm{Tr}\left(\mathcal{J}^a\mathcal{P}^b\right) = \eta^{ab}, \qquad \mathrm{Tr}\left(\mathcal{J}^a\mathcal{J}^b\right) = \mathrm{Tr}\left(\mathcal{P}^a\mathcal{P}^b\right) = 0, \tag{2.17}$$

then it is easy to verify that the action (2.1) is simply the Chern–Simons action [15] for A,

$$I_{\mathrm{CS}} = \frac{1}{2}\int_M \mathrm{Tr}\left\{A \wedge dA + \frac{2}{3}A \wedge A \wedge A\right\}. \tag{2.18}$$

The field equations now reduce to the requirement that A be a flat $ISO(2,1)$ connection, and the gauge transformations (2.2)–(2.3) can be identified with standard $ISO(2,1)$ gauge transformations. Imitating the arguments of our second interpretation, we should therefore expect solutions of the field equations to be characterized by gauge equivalence classes of flat $ISO(2,1)$ connections, that is by homomorphisms in the space

$$\tilde{\mathcal{M}} = \mathrm{Hom}(\pi_1(M), ISO(2,1))/\sim,$$
$$\rho_1 \sim \rho_2 \text{ if } \rho_2 = h \cdot \rho_1 \cdot h^{-1}, \quad h \in ISO(2,1). \tag{2.19}$$

To relate this description to our previous results, observe that $ISO(2,1)$ is itself a cotangent bundle with base space $SO(2,1)$. Indeed, a cotangent vector at the point $\Lambda_1 \in SO(2,1)$ can be written in the form $d\Lambda_1\Lambda_1^{-1}$, and the multiplication law

$$(\Lambda_1, d\Lambda_1\Lambda_1^{-1}) \cdot (\Lambda_2, d\Lambda_2\Lambda_2^{-1}) = (\Lambda_1\Lambda_2, d(\Lambda_1\Lambda_2)(\Lambda_1\Lambda_2)^{-1})$$
$$= (\Lambda_1\Lambda_2, d\Lambda_1\Lambda_1^{-1} + \Lambda_1(d\Lambda_2\Lambda_2^{-1})\Lambda_1^{-1}) \tag{2.20}$$

may be recognized as the standard semidirect product composition law for Poincaré transformations. The space of homomorphisms from $\pi_1(M)$ to $ISO(2,1)$ inherits this cotangent bundle structure in an obvious way, leading to the identification $\tilde{\mathcal{M}} \approx T^*\tilde{\mathcal{N}}$, where $\tilde{\mathcal{N}}$ is the space of homomorphisms (2.12). It remains for us to factor out the mapping class group. But this group acts in (2.12) and (2.19) as the same group of automorphisms of $\pi_1(M)$; writing the quotient as $\tilde{\mathcal{M}}/\sim' = \mathcal{M}$, we thus see that $\mathcal{M} \approx T^*\mathcal{N}$.

Of these four approaches to the (2+1)-dimensional field equations, only the first corresponds directly to our usual picture of spacetime physics. Trajectories of physical particles, for instance, are geodesics in the flat manifolds of this description. The second approach, on the other hand, is the one that is closest to the loop variable picture in (3+1)-dimensional gravity. The loop variables of Rovelli and Smolin [2, 3, 16, 17] may be expressed as follows. Let

$$U[\gamma, x] = P \exp \left(\int_\gamma \omega^a \mathcal{J}_a \right) \qquad (2.21)$$

be the holonomy of the connection 1-form ω^a around a closed path $\gamma(t)$ based at $\gamma(0) = x$. (Here, P denotes path ordering, and the basepoint x specifies the point at which the path ordering begins.) We then define

$$T^0[\gamma] = \text{Tr}\, U[\gamma, x] \qquad (2.22)$$

and

$$T^1[\gamma] = \int_\gamma dt\, \text{Tr} \left\{ U[\gamma, x(t)]\, e_\mu{}^a(\gamma(t)) \frac{dx^\mu}{dt}(\gamma(t)) \mathcal{J}_a \right\}. \qquad (2.23)$$

$T^0[\gamma]$ is thus the trace of the $SO(2,1)$ holonomy around γ, while $T^1[\gamma]$ is essentially a cotangent vector to $T^0[\gamma]$: indeed, given a curve $\omega(s)$ in the space of flat $SO(2,1)$ connections, we can differentiate (2.22) to obtain

$$\frac{d}{ds} T^0[\gamma] = \int_\gamma \text{Tr} \left(U[\gamma, x(t)] \frac{d\omega^a}{ds}(\gamma(t)) \mathcal{J}_a \right), \qquad (2.24)$$

and we have already seen that the derivative $d\omega^a/ds$ can be identified with the triad e^a.

In 3+1 dimensions, the variables T^0 and T^1 depend on particular loops γ, and considerable work is still needed to construct diffeomorphism-invariant observables that depend only on knot classes. In 2+1 dimensions, on the other hand, the loop variables are already invariant, at least under diffeomorphisms isotopic to the identity. The key difference is that in 2+1 dimensions the connection ω is flat, so the holonomy $U[\gamma, x]$ depends only on the homotopy class of γ. Some care must be taken in handling the mapping class group, which acts non-trivially on T^0 and T^1; this issue has been investigated in a slightly different context by Nelson and Regge [18].

Of course, T^0 and T^1 are not quite the equivalence classes of holonomies of our interpretation number 2 above: $T^0[\gamma]$ is not a holonomy, but only the *trace* of a holonomy. But knowledge of $T^0[\gamma]$ for a large enough set of homotopically inequivalent curves may be used to reconstruct a point in the space $\tilde{\mathcal{N}}$ of eqn (2.12), and indeed, the loop variables can serve as local coordinates on $\tilde{\mathcal{N}}$ [19, 20, 21].

3 Geometric structures: from holonomies to geometry

The central problem described in the introduction can now be made explicit. Spacetimes in 2+1 dimensions can be characterized à la Ashtekar *et al.* [16] as points in the cotangent bundle $T^*\mathcal{N}$, our description number 2 of the last section. Such a description is fully diffeomorphism invariant, and provides a natural starting point for quantization. But our intuitive geometric picture of a (2+1)-dimensional spacetime is that of a manifold M with a flat metric—description number 1—and only in this representation do we know how to connect the mathematics with ordinary physics. Our goal is therefore to provide a translation between these two descriptions.

To proceed, let us investigate the space of flat spacetimes in a bit more detail. If M is topologically trivial, the vanishing of the curvature tensor implies that (M, g) is simply ordinary Minkowski space $(V^{2,1}, \eta)$, or at least some subset of $(V^{2,1}, \eta)$ that can be extended to the whole of Minkowski space. If the spacetime topology is non-trivial, M can still be covered by contractible coordinate patches U_i that are each isometric to $V^{2,1}$, with the standard Minkowski metric $\eta_{\mu\nu}$ on each patch. The geometry is then encoded in the transition functions γ_{ij} on the intersections $U_i \cap U_j$, which determine how these patches are glued together. Moreover, since the metrics in U_i and U_j are identical, these transition functions must be isometries of $\eta_{\mu\nu}$, that is elements of the Poincaré group $ISO(2,1)$.

Such a construction is an example of what Thurston calls a geometric structure [22, 23, 24, 25], in this case a Lorentzian or $(ISO(2,1), V^{2,1})$ structure. In general, a (G, X) manifold is one that is locally modelled on X, just as an ordinary n-dimensional manifold is modelled on $\mathbf{R}^n$. More precisely, let G be a Lie group that acts analytically on some n-manifold X, the model space, and let M be another n-manifold. A (G, X) structure on M is then a set of coordinate patches U_i covering M with 'coordinates' $\phi_i : U_i \to X$ taking their values in the model space and with transition functions $\gamma_{ij} = \phi_i \circ \phi_j^{-1}|U_i \cap U_j$ in G. While this general formulation may not be widely known, specific examples are familiar: for example, the uniformization theorem for Riemann surfaces implies that any surface of genus $g > 1$ admits an $(PSL(2, \mathbf{R}), \mathbf{H}^2)$ structure.

A fundamental ingredient in the description of a (G, X) structure is its holonomy group, which can be viewed as a measure of the failure of a single coordinate patch to extend around a closed curve. Let M be a (G, X) manifold containing a closed path γ. We can cover γ with coordinate charts

$$\phi_i : U_i \to X, \qquad i = 1, \ldots, n \tag{3.1}$$

with constant transition functions $g_i \in G$ between U_i and U_{i+1}, i.e.

$$\phi_i|U_i \cap U_{i+1} = g_i \circ \phi_{i+1}|U_i \cap U_{i+1} \tag{3.2}$$

$$\phi_n | U_n \cap U_1 = g_n \circ \phi_1 | U_n \cap U_1.$$

Let us now try analytic continuation of the coordinate ϕ_1 from the patch U_1 to the whole of γ. We can begin with a coordinate transformation in U_2 that replaces ϕ_2 by $\phi_2' = g_1 \circ \phi_2$, thus extending ϕ_1 to $U_1 \cup U_2$. Continuing this process along the curve, with $\phi_j' = g_1 \circ \ldots \circ g_{j-1} \circ \phi_j$, we will eventually reach the final patch U_n, which again overlaps U_1. If the new coordinate function $\phi_n' = g_1 \circ \ldots \circ g_{n-1} \circ \phi_n$ happens to agree with ϕ_1 on $U_n \cap U_1$, we will have succeeded in covering γ with a single patch. Otherwise, the holonomy H, defined as $H(\gamma) = g_1 \circ \ldots \circ g_n$, measures the obstruction to such a covering.

It may be shown that the holonomy of a curve γ depends only on its homotopy class [22]. In fact, the holonomy defines a homomorphism

$$H : \pi_1(M) \to G. \tag{3.3}$$

Note that if we pass from M to its universal covering space $\widetilde{M}$, we will no longer have non-contractible closed paths, and ϕ_1 will be extendable to all of $\widetilde{M}$. The resulting map $D : \widetilde{M} \to X$ is called the developing map of the (G, X) structure. At least in simple examples, D embodies the classical geometric picture of development as 'unrolling'—for instance, the unwrapping of a cylinder into an infinite strip.

The homomorphism H is not quite uniquely determined by the geometric structure, since we are free to act on the model space X by a fixed element $h \in G$, thus changing the transition functions g_i without altering the (G, X) structure of M. It is easy to see that such a transformation has the effect of conjugating H by h, and it is not hard to prove that H is in fact unique up to such conjugation [22]. For the case of a Lorentzian structure, where $G = ISO(2, 1)$, we are thus led to a space of holonomies of precisely the form (2.19).

This identification is not a coincidence. Given a (G, X) structure on a manifold M, it is straightforward to define a corresponding flat G bundle [24]. To do so, we simply form the product $G \times U_i$ in each patch—giving the local structure of a G bundle—and use the transition functions γ_{ij} of the geometric structure to glue together the fibers on the overlaps. It is then easy to verify that the flat connection on the resulting bundle has a holonomy group isomorphic to the holonomy group of the geometric structure.

We can now try to reverse this process, and use one of the holonomy groups of eqn (2.19)—approach number 4 to the field equations—to define a Lorentzian structure on M, reproducing approach number 1. In general, this step may fail: the holonomy group of a (G, X) structure is not necessarily sufficient to determine the full geometry. For spacetimes, it is easy to see what can go wrong. If we start with a flat 3-manifold M and simply cut out a ball, we can obtain a new flat manifold without affecting the holonomy of the geometric structure. This is a rather trivial change, however, and we would like to show that nothing worse can go wrong.

Mess [9] has investigated this question for the case of spacetimes with topologies of the form $\mathbf{R} \times \Sigma$. He shows that the holonomy group determines a unique 'maximal' spacetime M—specifically, a spacetime constructed as a domain of dependence of a spacelike surface Σ. Mess also demonstrates that the holonomy group H acts properly discontinuously on a region $W \subset V^{2,1}$ of Minkowski space, and that M can be obtained as the quotient space W/H. This quotient construction can be a powerful tool for obtaining a description of M in reasonably standard coordinates, for instance in a time slicing by surfaces of constant mean curvature.

For topologies more complicated than $\mathbf{R} \times \Sigma$, I know of very few general results. But again, a theorem of Mess is relevant: if M is a compact 3-manifold with a flat, non-degenerate, time-orientable Lorentzian metric and a strictly spacelike boundary, then M necessarily has the topology $\mathbf{R} \times \Sigma$, where Σ is a closed surface homeomorphic to one of the boundary components of M. This means that for spatially closed 3-dimensional universes, topology change is classically forbidden, and the full topology is uniquely fixed by that of an initial spacelike slice. Hence, although more exotic topologies may occur in some approaches to quantum gravity, it is not physically unreasonable to restrict our attention to spacetimes $\mathbf{R} \times \Sigma$.

To summarize, we now have a procedure—valid at least for spacetimes of the form $\mathbf{R} \times \Sigma$—for obtaining a flat geometry from the invariant data given by Ashtekar–Rovelli–Smolin loop variables. First, we use the loop variables to determine a point in the cotangent bundle $T^*\mathcal{N}$, establishing a connection to our second approach to the field equations. Next, we associate that point with an $ISO(2,1)$ holonomy group $H \in \mathcal{M}$, as in our approach number 4. Finally, we identify the group H with the holonomy group of a Lorentzian structure on M, thus determining a flat spacetime of approach number 1. In particular, if we can solve the (difficult) technical problem of finding an appropriate fundamental region $W \subset V^{2,1}$ for the action of H, we can write M as a quotient space W/H.

This procedure has been investigated in detail for the case of a torus universe, $\mathbf{R} \times T^2$, in [26] and [11]. For a universe containing point particles, it is implicit in the early descriptions of Deser *et al.* [27], and is explored in some detail in [10]. For the (2+1)-dimensional black hole, the geometric structure can be read off from [28] and [29].[12] And although it is never stated explicitly, the recent work of 't Hooft [31] and Waelbroeck [32] is really a description of flat spacetimes in terms of Lorentzian structures.

---

[12]For the black hole, a cosmological constant must be added to the field equations. Instead of being flat, the resulting spacetime has constant negative curvature, and the geometric structure becomes an $(SO(2,2), \mathbf{H}^{2,1})$ structure. A related result for the torus will appear in [30].

4   Quantization and geometrical observables

Our discussion so far has been strictly classical. I would like to conclude by briefly describing some of the issues that arise if we attempt to quantize (2+1)-dimensional gravity.

The canonical quantization of a classical system is by no means uniquely defined, but most approaches have some basic features in common. A classical system is characterized by its phase space, a $2N$-dimensional symplectic manifold Γ, with local coordinates consisting of N position variables and N conjugate momenta. Classical observables are functions of the positions and momenta, that is maps $f, g, \ldots$ from Γ to $\mathbf{R}$. The symplectic form Ω on Γ determines a set of Poisson brackets $\{f, g\}$ among observables, and hence induces a Lie algebra structure on the space of observables. To quantize such a system, we are instructed to replace the classical observables with operators and the Poisson brackets with commutators; that is, we are to look for an irreducible representation of this Lie algebra as an algebra of operators acting on some Hilbert space.

As stated, this program cannot be carried out: Van Hove showed in 1951 that in general, no such irreducible representation of the full Poisson algebra of classical observables exists [33]. In practice, we must therefore choose a subalgebra of 'preferred' observables to quantize, one that must be small enough to permit a consistent representation and yet big enough to generate a large class of classical observables [34]. Ordinarily, the resulting quantum theory will depend on this choice of preferred observables, and we will have to look hard for physical and mathematical justifications for our selection.

In simple classical systems, there is often an obvious set of preferred observables—the positions and momenta of point particles, for instance, or the fields and their canonical momenta in a free field theory. For gravity, on the other hand, such a natural choice seems difficult to find. In 2+1 dimensions, where a number of approaches to quantization can be carried out explicitly, it is known that different choices of variables lead to genuinely different quantum theories [35, 36, 37].

In particular, each of the four interpretations of the field equations discussed above suggests its own set of fundamental observables. In the first interpretation—solutions as flat spacetimes—the natural candidates are the metric and its canonical momentum on some spacelike surface. But these quantities are not diffeomorphism invariant, and it seems that the best we can do is to define a quantum theory in some particular, fixed time slicing [26, 38, 39]. This is a rather undesirable situation, however, since the choice of such a slicing is arbitrary, and there is no reason to expect the quantum theories coming from different slicings to be equivalent.

In our second interpretation—solutions as classes of flat connections and their cotangents—the natural observables are points in the bundle $T^*\mathcal{N}$.

These are diffeomorphism invariant, and quantization is relatively straight-forward; in particular, the appropriate symplectic structure for quantization is just the natural symplectic structure of $T^*\mathcal{N}$ as a cotangent bundle. The procedure for quantizing such a cotangent bundle is well established [2], and there seem to be no fundamental difficulties in constructing the quantum theory. But now, just as in the classical theory, the physical interpretation of the quantum observables is obscure.

It is therefore natural to ask whether we can extend the classical relationships between these approaches to the quantum theories. At least for simple topologies, the answer is positive. The basic strategy is as follows.

We begin by choosing a set of physically interesting classical observables of flat spacetimes. For example, it is often possible to foliate uniquely a spacetime by spacelike hypersurfaces of constant mean extrinsic curvature $\mathrm{Tr}K$; the intrinsic and extrinsic geometries of such slices are useful observables with clear physical interpretations. Let us denote these variables generically as $Q(T)$, where the parameter T labels the time slice on which the Q are defined (for instance, $T = \mathrm{Tr}K$).

Classically, such observables can be determined—at least in principle—as functions of the geometric structure, and thus of the $ISO(2,1)$ holonomies ρ,

$$Q = Q[\rho, T]. \tag{4.1}$$

We now adopt these holonomies as our preferred observables for quantization, obtaining a Hilbert space $L^2(\mathcal{N})$ and a set of operators $\hat{\rho}$. Finally, we translate (4.1) into an operator equation,

$$\widehat{Q} = \widehat{Q}[\hat{\rho}, T], \tag{4.2}$$

thus obtaining a set of diffeomorphism-invariant but 'time-dependent' quantum observables to represent the variables Q. Some ambiguity will remain, since the operator ordering in (4.2) is rarely unique, but in the examples studied so far, the requirement of mapping class group invariance seems to place major restrictions on the possible orderings [35].

This program has been investigated in some detail for the simplest non-trivial topology, $M = \mathbf{R} \times T^2$ [26, 35]. There, a natural set of 'geometric' variables are the modulus τ of a toroidal slice of constant mean curvature $\mathrm{Tr}K = T$ and its conjugate momentum p_τ. These can be expressed explicitly in terms of a set of loop variables that characterize the $ISO(2,1)$ holonomies of the spacetime. Following the program outlined above, one obtains a 1-parameter family of diffeomorphism-invariant operators $\hat{\tau}(T)$ and $\hat{p}_\tau(T)$ that describe the physical evolution of a spacelike slice. The T-dependence of these operators can be described by a set of Heisenberg equations of motion,

$$\frac{d\hat{\tau}}{dT} = i\left[\hat{H}, \hat{\tau}\right], \qquad \frac{d\hat{p}_\tau}{dT} = i\left[\hat{H}, \hat{p}_\tau\right], \qquad (4.3)$$

with a Hamiltonian $\hat{H}[\hat{\tau}, \hat{p}_\tau, T]$ that can again be calculated explicitly.

Let me stress that despite the familiar appearance of (4.3), the parameter T is *not* a time coordinate in the ordinary sense; the operators $\hat{\tau}(T)$ and $\hat{p}_\tau(T)$ are fully diffeomorphism invariant. We thus have a kind of 'time dependence without time dependence', an expression of dynamics in terms of operators that are individually constants of motion. For more complicated topologies, such an explicit construction seems quite difficult, although Unruh and Newbury have taken some interesting steps in that direction [40]. Ideally, one would like to find some kind of perturbation theory for geometrical variables like $\widehat{Q}$, but little progress has yet been made in this direction.

The specific constructions I have described here are unique to 2+1 dimensions, of course. But I believe that some of the basic features are likely to extend to realistic (3+1)-dimensional gravity. The quantization of holonomies is a form of 'covariant canonical quantization' [41, 42], or quantization of the space of classical solutions. We do not yet understand the classical solutions of the (3+1)-dimensional field equations well enough to duplicate such a strategy, but a similar approach may be useful in minisuperspace models. The invariant but T-dependent operators $\hat{\tau}(T)$ and $\hat{p}_\tau(T)$ are examples of Rovelli's 'evolving constants of motion' [43], whose use has also been suggested in (3+1)-dimensional gravity. Finally, our simple model has strikingly confirmed the power of the Rovelli–Smolin loop variables. A full extension to 3+1 dimensions undoubtedly remains a distant goal, but for the first time in years, there seems to be some real cause for optimism.

Acknowledgements

This work was supported in part by US Department of Energy grant DE-FG03-91ER40674.

Bibliography

1. A. Ashtekar, *Phys. Rev.* **D36** (1987) 1587.

2. A. Ashtekar, *Lectures on Nonperturbative Quantum Gravity* (World Scientific, Singapore, 1991).

3. C. Rovelli and L. Smolin, *Nucl. Phys.* **B331** (1990) 80.

4. For a recent review, see L. Smolin, in *General Relativity and Gravitation 1992*, ed. R. J. Gleiser, C. N. Kozameh, and O. M. Moreschi (IOP Publishing, Bristol, 1993).

5. A. Ashtekar, C. Rovelli, and L. Smolin, *Phys. Rev. Lett.* **69** (1992)

237.

6. E. Witten, *Nucl. Phys.* **B311** (1988) 46.

7. W. J. Harvey, in *Discrete Groups and Automorphic Functions*, ed. W. J. Harvey (Academic Press, New York, 1977).

8. W. M. Goldman, *Invent. Math.* **93** (1988) 557.

9. G. Mess, 'Lorentz Spacetimes of Constant Curvature', Institut des Hautes Etudes Scientifiques preprint IHES/M/90/28 (1990).

10. S. Carlip, *Classical & Quantum Gravity* **8** (1991) 5.

11. J. Luoko and D. M. Marolf, 'Solution Space to 2+1 Gravity on $\mathbf{R} \times T^2$ in Witten's Connection Formulation', Syracuse preprint SU-GP-93/7-6 (1993).

12. W. M. Goldman, *Adv. Math.* **54** (1984) 200.

13. L. Smolin, in *Knots, Topology, and Quantum Field Theory*, ed. L. Lusanna (World Scientific, Singapore, 1989).

14. For an extensive review, see K. Kuchař, in *Proc. of the 4th Canadian Conf. on General Relativity and Relativistic Astrophysics*, ed. G. Kunstatter *et al.* (World Scientific, Singapore, 1992).

15. E. Witten, *Commun. Math. Phys.* **121** (1989) 351.

16. A. Ashtekar, V. Husein, C. Rovelli, J. Samuel, and L. Smolin, *Classical & Quantum Gravity* **6** (1989) L185.

17. See also T. Jacobson and L. Smolin, *Nucl. Phys.* **B299** (1988) 295.

18. J. E. Nelson and T. Regge, *Commun. Math. Phys.* **141** (1991) 211; *Phys. Lett.* **B272** (1991) 213.

19. M. Seppälä and T. Sorvali, *Ann. Acad. Sci. Fenn. Ser. A. I. Math.* **10** (1985) 515.

20. T. Okai, *Hiroshima Math. J.* **22** (1992) 259.

21. J. E. Nelson and T. Regge, *Commun. Math. Phys.* **155** (1993) 561.

22. W.P. Thurston, *The Geometry and Topology of Three-Manifolds*, Princeton lecture notes (1979).

23. R. D. Canary, D. B. A. Epstein, and P. Green, in *Analytical and Geometric Aspects of Hyperbolic Space*, London Math. Soc. Lecture Notes Series **111**, ed. D. B. Epstein (Cambridge University Press, Cambridge, 1987).

24. W. M. Goldman, in *Geometry of Group Representations*, ed. W. M. Goldman and A. R. Magid (American Mathematical Society, Providence, 1988).

25. For examples of geometric structures, see D. Sullivan and W. Thurston, *Enseign. Math.* **29** (1983) 15.

26. S. Carlip, *Phys. Rev.* **D42** (1990) 2647.

27. S. Deser, R. Jackiw, and G. 't Hooft, *Ann. Phys.* **152** (1984) 220.

28. M. Bañados *et al.*, *Phys. Rev.* **D48** (1993) 1506.

29. D. Cangemi, M. Leblanc, and R. B. Mann, 'Gauge Formulation of the Spinning Black Hole in (2+1)-Dimensional Anti-de Sitter Space', MIT preprint MIT-CTP-2162 (1992).

30. S. Carlip and J. E. Nelson, 'Equivalent Quantisations of (2+1)-Dimensional Gravity', Turin preprint DFTT 67/93 and Davis preprint UCD-93-33.

31. G. 't Hooft, *Classical & Quantum Gravity* **9** (1992) 1335; *Classical & Quantum Gravity* **10** (1993) 1023.

32. H. Waelbroeck, *Classical & Quantum Gravity* **7** (1990) 751; *Phys. Rev. Lett.* **64** (1990) 2222; *Nucl. Phys.* **B364** (1991) 475.

33. L. Van Hove, *Mem. de l'Acad. Roy. de Belgique (Classe de Sci.)* **37** (1951) 610.

34. For a useful summary, see C. J. Isham, in *Relativity, Groups, and Topology II*, ed. B. S. DeWitt and R. Stora (North-Holland, Amsterdam, 1984).

35. S. Carlip, *Phys. Rev.* **D45** (1992) 3584; *Phys. Rev.* **D47** (1993) 4520.

36. S. Carlip, 'Six Ways to Quantize (2+1)-Dimensional Gravity', Davis preprint UCD-93-15, to appear in *Proc. of the Fifth Canadian Conference on General Relativity and Relativistic Astrophysics*.

37. D. M. Marolf, 'Loop Representations for 2+1 Gravity on a Torus', Syracuse preprint SU-GP-93/3-1 (1993).

38. V. Moncrief, *J. Math. Phys.* **30** (1989) 2907.

39. A. Hosoya and K. Nakao, *Classical & Quantum Gravity* **7** (1990) 163.

40. W. G. Unruh and P. Newbury, 'Solution to 2+1 Gravity in the Dreibein Formalism', University of British Columbia preprint (1993).

41. A. Ashtekar and A. Magnon, *Proc. R. Soc. (London)* **A346** (1975) 375.

42. C. Crnkovic and E. Witten, in *Three Hundred Years of Gravity*, ed. S. W. Hawking and W. Israel (Cambridge University Press, Cambridge, 1987).

43. C. Rovelli, *Phys. Rev.* **D42** (1990) 2638; *Phys. Rev.* **D43** (1991) 442.

From Chern–Simons to WZW via Path Integrals

Dana S. Fine

Department of Mathematics, University of Massachusetts,
Dartmouth, Massachusetts 02747, USA
(email: dfine@cis.umassd.edu)

Abstract

Direct analysis of the Chern–Simons path integral reduces quantum expectations of unknotted Wilson lines in Chern–Simons theory on S^3 with group G to n-point functions in the WZW model of maps from S^2 to G. The reduction hinges on the characterization of $\mathcal{A}/\mathcal{G}_n$, the space of connections modulo those gauge transformations which are the identity at a point n, as itself a principal fiber bundle with affine-linear fiber.

1 Introduction

In this chapter, I will describe how to reduce the path integral for Chern–Simons theory on S^3 to the path integral for the Wess–Zumino–Witten (WZW) model for maps from S^2 to G, where G is the symmetry group of the Chern–Simons theory. The points I would like to emphasize are, first, that the Chern–Simons path integral on S^3 is tractable, and, second, that it provides an unusually direct link between the Chern–Simons theory and the WZW model. The details I omit appear in [4].

That the two theories are related has been known at least since Witten first described the geometric quantization of the Chern–Simons theory in the axial gauge on manifolds of the form $R \times \Sigma$, where Σ is a Riemann surface [9]. Although this is frequently described as a path integral approach, it does not use a direct evaluation of the path integral; rather, the formal properties of the path integral provide axioms crucial to extending the theory from $R \times \Sigma$ to compact 3-manifolds. In this approach, the WZW model appears in a highly refined form, as the source of a certain modular functor. That is, one must already know how to *solve* the WZW model (or at least know that its solution defines this modular functor) to recognize its appearance in the Chern–Simons theory.

Treatments of the path integral for Chern–Simons in the axial gauge on $R \times \Sigma$ tend to make contact with the WZW model at earlier stages. Moore and Seiberg [6] proceed with the evaluation of the path integral far enough to obtain a path integral over the phase space LG/G. This links the two theories as theories of the representations of LG/G. Fröhlich and King [5] obtain the Knizhnik–Zamolodchikov equation of the WZW model from the same gauge-fixed Chern–Simons path integral.

The approach I will describe is novel in that it treats the path integral on S^3 directly, it does not require a choice of gauge, and one need only know the action of the WZW model, not its solution nor even its current algebra, to recognize its appearance in the Chern–Simons theory. The key ingredient is a technique for integrating directly on the space of connections modulo gauge transformations which I developed to evaluate path integrals for Yang–Mills on Riemann surfaces [3].

2   The main result

The Chern–Simons path integral to evaluate is

$$\langle f \rangle_{CS} = \frac{1}{Z_0} \int_{\mathcal{A}/\mathcal{G}_n} f([A]) e^{iS(A)} \rho_*(\mathcal{D}A),$$

where A is a connection on the principal bundle $P = S^3 \times G$, and $S(A)$ is the Chern–Simons action:

$$\frac{k}{4\pi} \int_{S^3} \text{Tr}\left(A \wedge dA + \frac{1}{3} A \wedge [A \wedge A] \right).$$

Here $\mathcal{A}$ is the space of smooth connections on P, $\mathcal{G}_n$ is the space of gauge transformations which are the identity at the north pole n of S^3, $\rho : \mathcal{A} \to \mathcal{A}/\mathcal{G}_n$ is the standard map to the quotient, $\mathcal{D}A$ means the standard measure on $\mathcal{A}$, and $\rho_*(\mathcal{D}A)$ is the push-forward to $\mathcal{A}/\mathcal{G}_n$. Note that even the formal definition of $\mathcal{D}A$ requires a metric. Take this to be the standard metric on S^3. The WZW path integral for $\Omega^2 G$ (based maps from S^2 to G) is

$$\langle f \rangle_{WZW} = \frac{1}{Z_0} \int_{\Omega^2 G} f(X) e^{iS(X)} \mathcal{D}X,$$

where the WZW action is

$$S(X) = \frac{k}{4\pi} \left(\int_{S^2} \text{Tr}(\partial X \wedge \bar{\partial} X) + \frac{1}{3} \int_M \text{Tr}(\widehat{X}^{-1} d\widehat{X} \wedge [\widehat{X}^{-1} d\widehat{X} \wedge \widehat{X}^{-1} d\widehat{X}]) \right).$$

Here M is any 3-manifold with $\partial M = S^2$, and $\widehat{X}$ is any extension of X from S^2 to M.

The main result is

$$\langle f \rangle_{CS} = \frac{1}{Z_0} \int \hat{f}(X) e^{i\frac{k}{4\pi} \int_{S^2} \text{Tr}(\partial X \wedge \bar{\partial} X) + i\frac{k}{12\pi} \int_M \text{Tr}(\widehat{X}^{-1} d\widehat{X})^3} \mathcal{D}X,$$

where $X \in \Omega^2 G$, and f explicitly determines $\hat{f}$. In particular, if f is a Wilson line, then $\hat{f}$ is the trace of a product of evaluation-at-a-point

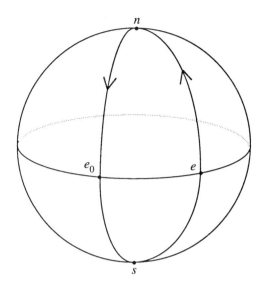

FIG. 1. The path $\gamma(e)$

maps. That is, the Chern–Simons path integral directly reduces to the path integral for a WZW model of based maps from S^2 to G.

The technique is as follows:

1. Characterize $\mathcal{A}/\mathcal{G}_n$ as a bundle over $\Omega^2 G$ with affine-linear fibers.
2. Integrate along the fiber over each point $X \in \Omega^2 G$.
3. Recognize the resulting 'measure' on $\Omega^2 G$ as a WZW path integral.

3 $\mathcal{A}/\mathcal{G}_n$ as a bundle over $\Omega^2 G$

For each connection A, and each point $e \in S^2$, define $\xi_A(e)$ by holonomy about paths $\gamma(e)$ of the form pictured in Fig. 1. That is, $\gamma(e)$ follows a longitude from the north pole n to the south pole s through a fixed point e_0 of the equator and then returns along the longitude through the (variable) point e. Relative to a basepoint in the fiber over n, $\xi_A(e)$ lies in G. As e varies throughout the equator (identified with S^2), $\xi_A(e)$ defines, for each A, a map from S^2 to G. Since $\xi_A(e_0) = \mathbf{1}$, it is a based map, so $\xi_A \in \Omega^2 G$. Moreover, since a gauge transformation affects holonomy about $\gamma(e)$ by conjugation by the value of the gauge transformation at n, $\xi_{\phi \bullet A} = \xi_A$ for $\phi \in \mathcal{G}_n$. Thus the assignment $[A] \mapsto \xi_A$ defines a mapping $\xi : \mathcal{A}/\mathcal{G}_n \to \Omega^2 G$ which will serve as our projection.

Turn now to the fiber of ξ. Clearly, $\xi_{A+\tau} = \xi_A$ if τ vanishes along the longitudes of Fig. 1. If $P_0 : \bigwedge^1(S^3, \mathbf{g}) \to \bigwedge^1(S^3, \mathbf{g})$ denotes projection to the longitudinal part, the space of such τ is $\ker P_0$. In fact, the fiber over

ξ_A is the set of orbits

$$\{[A + \tau] : \tau \in \ker P_0\}. \tag{3.1}$$

Thus the fiber is the affine version of the linear space $\ker P_0$, and the connection A in the above presentation represents a choice of origin in the fiber.

The image of ξ is all of $\Omega^2 G$. To see this, use a homotopy from X to the identity to determine lifts to P of the family of paths $\gamma(e)$ in S^3 which defined ξ. (Such a homotopy exists, because $\pi_2(G) = 0$ for any Lie group G). This determines $P_0 A$ such that $\xi(A) = X$. This is a slight refinement of [7] and [1], which describe a homotopy equivalence between $\mathcal{A}/\mathcal{G}_n$ for bundles over S^4 and $\Omega^3 G$. The reason to redo this is to describe the (topologically trivial) fiber in an explicit manner.

According to eqn 3.1, the integral over the fiber looks like

$$I_{\text{fiber}}(X) = \det{}^{1/2}(D_A^* P_0 D_A) \int f([A + \tau]) e^{iS(A+\tau)} \, \mathcal{D}\tau.$$

The determinant factor relates $\rho_*(\mathcal{D}A)$ on the fiber to the metric-compatible measure $\mathcal{D}\tau$. The action along the fiber is, after an integration by parts,

$$S(A + \tau) = S(A) + \frac{k}{4\pi} \left[\int_M \text{Tr}\left(\tau \wedge D_A \tau + 2\, F_A \wedge \tau\right) + \int_{\partial M} \text{Tr}(A \wedge \tau) \right],$$

where, for reasons which will soon be clear, the boundary contribution is retained. To integrate over the fiber, choose a specific origin $[\tilde{A}]$ to remove the linear term.

One such choice is determined by

$$\tilde{A} = \widehat{X}^{-1} d\widehat{X},$$

where $\widehat{X} : S^3 \to G$ is parallel transport by A along the longitudinal paths $\gamma(e)$ used in defining ξ. Note that $\widehat{X}$ is not well defined at n. In fact $\widehat{X}|_\partial = X$, where $|_\partial$ is the restriction to a small 2-sphere approaching n. Thus $\tilde{A}$ is not continuous at n (nor is any other such choice), so to keep $\tilde{A} + \tau$ smooth requires

$$\tau|_\partial = - \left.\tilde{A}\right|_\partial = -X^{-1} dX.$$

With this choice, the boundary term vanishes, and

$$S(\tilde{A} + \tau) = S(\tilde{A}) + \frac{k}{4\pi} \int_{S^3} \text{Tr}(\tau \wedge D_A \tau).$$

The integral over the fiber becomes

$$I_{\text{fiber}}(X) = \det{}^{1/2}(D_A^* P_0 D_A) e^{iS(\tilde{A})} \int_{\widehat{\ker P_0}} f(\tilde{A} + \tau) e^{i\frac{k}{4\pi} \int \text{Tr}(\tau \wedge D_A \tau)} \, \mathcal{D}\tau,$$

where $\widehat{\ker P_0}$ denotes elements of $\ker P_0$ with the prescribed discontinuity at n. Now assume f is independent of τ and consider $\int e^{i\frac{k}{4\pi}\int \mathrm{Tr}(\tau \wedge D_A\tau)}\, \mathcal{D}\tau$. But for the discontinuity at n, this would be standard. The result would be a determinant times the exponential of the eta-invariant as in [8] and [9]. Repeating these arguments, which refer to a finite-dimensional analog, with a minor variation to account for the discontinuity, will evaluate I_{fiber}. The metric on $\bigwedge^1(S^3, \mathbf{g})$ restricts to $\ker P_0$ to define a decomposition (of its complexification) $\ker P_0^C = \bigwedge_+ \oplus \bigwedge_-$, in terms of which

$$\int \mathrm{Tr}(\tau \wedge D_A\tau) = 2\int_{S^3}\mathrm{Tr}(\tau_+ \wedge D_A\tau_-) - \int_{S^2}\mathrm{Tr}(\tau_- \wedge \tau_+)$$

$$= 2\int_{S^3}\mathrm{Tr}(\tau_+ \wedge D_A\tau_-) + \int_{S^2}\mathrm{Tr}(\partial X \wedge \bar{\partial}X).$$

Model this by a vector space $V = V_+ \oplus V_-$ with a linear mapping $T: V_\pm \to (V_\mp)^*$. Add a small quadratic term, and integrate to find

$$\int e^{2i\langle Tv_-, v_+\rangle}\, \mathcal{D}v_-\mathcal{D}v_+ = \det{}^{-1/2}\left(T_-^*T_-\right).$$

The determinant analogous to the right-hand side is that of $P_-D_A^*D_AP_-$ acting on elements of $\ker P_0$ which vanish at n. Now, regularize and account for negative eigenvalues using the eta-invariant; that is, take

$$\det{}^{-\frac{1}{2}}|P_-D_A^*D_AP_-|\,e^{-i\frac{\pi}{4}\eta(0)},$$

where $\eta(0)$ is the eta-invariant, as the analogue for $\det^{-1/2}\left(T_-^*T_-\right)$. Then

$$I_{\mathrm{fiber}}(X) = \frac{\det^{\frac{1}{2}}\left(D_A^*P_0D_A\right)}{\det^{\frac{1}{2}}|P_-D_A^*D_AP_-|}e^{-i\frac{\pi}{4}\eta(0)}e^{iS(\tilde{A})}e^{i\frac{k}{4\pi}\int \mathrm{Tr}(\partial X \wedge \bar{\partial}X)}.$$

Performing the integration over the fibers in the path integral for $\langle f \rangle$, assuming f constant along the fibers, thus gives

$$\langle f \rangle = \frac{1}{Z_0}\int f(\tilde{A})e^{iS(\tilde{A})}e^{i\frac{k}{4\pi}\int \mathrm{Tr}(\partial X \wedge \bar{\partial}X)}\frac{\det^{\frac{1}{2}}\left(D_A^*P_0D_A\right)}{\det^{\frac{1}{2}}|P_-D_A^*D_AP_-|}e^{-i\frac{\pi}{4}\eta(0)}\mu_{\mathrm{base}},$$

where μ_{base} is the measure on $\Omega^2 G$ induced by the metric on $\mathcal{A}/\mathcal{G}_n$. To interpret this in terms of $\Omega^2 G$ requires four observations:

1. $S\left(\tilde{A}\right) = \frac{k}{12\pi}\int_{S^3}\mathrm{Tr}\left(\widehat{X}^{-1}d\widehat{X}\right)^3.$

2. $\mu_{\mathrm{base}} = \text{constant} \times \mathcal{D}X.$

3. Each determinant is constant along the *base* in $\mathcal{A}/\mathcal{G}_n$.

4. f can be expressed as a function $\hat{f}$ of X.

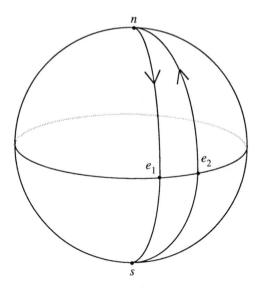

FIG. 2. The path whose holonomy is $X(e_1)^{-1}X(e_2)$

Absorbing everything independent of X into Z_0, then,

$$\langle f \rangle_{\mathrm{CS}} = \frac{1}{Z_0} \int \hat{f}(X) e^{i\frac{k}{4\pi}\int_{S^2}\mathrm{Tr}(\partial X\wedge\bar{\partial}X)+i\frac{k}{12\pi}\int_{S^3}\mathrm{Tr}\left(\hat{X}^{-1}d\hat{X}\right)^3} \mathcal{D}X,$$

as claimed.

4   An explicit reduction from Chern–Simons to WZW

Let R be a representation of G, and let $\mathcal{W}_R$ denote the Wilson line corresponding to the longitudinal path indicated in Fig. 2. Then

$$\langle \mathcal{W}_R \rangle_{\mathrm{CS}} =$$

$$\frac{1}{Z_0} \int \mathrm{Tr}_R \left(X(e_1)^{-1}X(e_2) \right) e^{i\frac{k}{4\pi}\int \mathrm{Tr}(\partial X\wedge\bar{\partial}X)+i\frac{k}{12\pi}\int_{S^3}\mathrm{Tr}\left(\hat{X}^{-1}d\hat{X}\right)^3} \mathcal{D}X$$

$$= \left\langle \mathrm{Tr}_R \left(X(e_1)^{-1}X(e_2) \right) \right\rangle_{\mathrm{WZW}}.$$

As a description of the relation between Chern–Simons theory on S^3 and the WZW for $\Omega^2 G$, this work is nearly complete. However, as an evaluation of the Chern–Simons path integral it lacks an essential component; namely, an evaluation of the WZW path integral appearing on the right-hand side of the above equation. For expectations of evaluation-at-a-point maps, the WZW path integral is tractable, at least formally, but this is work in progress. The basic idea is to reinterpret $X \in \Omega^2 G$ as an element of $\mathrm{Path}(\Omega G)$ with some constraints on its endpoint values. Then, the WZW path integral becomes a path integral for quantum mechanics on ΩG. At a formal level, expectations of evaluation-at-a-point maps are given by the Feynman–Kac formula. Formal calculations lead to some promising preliminary observations such as ⟨complicated unknot⟩ = ⟨unknot⟩. However, to come up with an explicit evaluation of even ⟨unknot⟩ requires going beyond formal properties to a more detailed understanding of the heat kernel on ΩG.

Acknowledgements

This material is based upon work supported in part by the National Science Foundation under Grant # DMS-9307608.

Bibliography

1. M. F. Atiyah and J. D. S. Jones, Topological aspects of Yang–Mills theory, *Commun. Math. Phys.* **61** 97–118 (1978)

2. D. Fine, Yang–Mills on the two-sphere, *Commun. Math. Phys.* **134** 273–292 (1990)

3. D. Fine, Yang-Mills on a Riemann surface, *Commun. Math. Phys.* **140**, 321–338 (1991)

4. D. Fine, Chern–Simons on S^3 via path integrals, preprint

5. J. Fröhlich and C. King, The Chern–Simons theory and knot polynomials, *Commun. Math. Phys.* **126** 167–199 (1989)

6. G. Moore and N. Seiberg, Taming the conformal zoo, *Phys. Lett.* **B220** 422–430 (1989)

7. I. M. Singer, Some remarks on the Gribov ambiguity, *Commun. Math. Phys.* **60** 7 (1978)

8. I. M. Singer, Families of Dirac operators with applications to physics, *Société Mathématique de France Astérisque*, hor série, 323–340 (1985)

9. E. Witten, Quantum field theory and the Jones polynomial, *Commun. Math. Phys.* **121** 351–399 (1989)

Topological Field Theory as the Key to Quantum Gravity

Louis Crane

Mathematics Department, Kansas State University,
Manhattan, Kansas 66506, USA
(email: crane@math.ksu.edu)

Abstract

Motivated by the similarity between CSW theory and the Chern–Simons state for general relativity in the Ashtekar variables, we explore what a universe would look like if it were in a state corresponding to a 3d TQFT. We end up with a construction of propagating states for parts of the universe and a Hilbert space corresponding to a certain approximation. The construction avoids path integrals, using instead recombination diagrams in a very special tensor category.

1 Introduction

What I wish to propose is that the quantum theory of gravity can be constructed from a topological field theory. Notice, I do not say that it **is** a topological field theory. Physicists will be quick to point out that there are local excitations in gravity, so it cannot be topological.

Nevertheless, the connections between general relativity in the loop formulation and the CSW TQFT are very suggestive. Formally, at least, the CSW action term

$$e^{\frac{ik}{4\pi}\int \mathrm{Tr}(A \wedge dA + \frac{2}{3} A \wedge A \wedge A)} \tag{1.1}$$

gives a state for the Ashtekar variable version of quantized general relativity, as mentioned previously in this volume. That is, we can think of this 'measure' on the space of connections on a 3-manifold either as defining a theory in 3 dimensions, or as a state in a 4d theory, which is GR with a cosmological constant.

Furthermore, states for the Ashtekar or loop variable form of GR are very scarce, unless one is prepared to consider states of zero volume. Finally, one should note that a number of very talented workers in the field have searched in vain for a Hilbert space of states in either representation.

If one attempts to use the loop variables to quantize subsystems of the universe, one ends up looking at functionals on the set of links in a 3-manifold with boundary with ends on the boundary. It turns out in regularizing that the links need to be framed, and one can trace them in any representation of $SU(2)$, so they need to be labelled. The coincidence of all this with the picture of 3d TQFT was the initial motivation for this work.

Another point, which seems important, is that although path integrals in general do not exist, the CSW path integral can be thought of as a symbolic shorthand for a 3d TQFT which can be rigorously defined by combining combinatorial topology with some very new algebraic structures, called modular tensor categories. Thus, if we want to explore (1.1) as a state for GR, we have the possibility of switching from the language of path integrals to a finite, discrete picture, where physics is described by variables on the edges of a triangulation.

Physics on a lattice is nothing new. What is new here is that the choice of lattice is unimportant, because the coefficients attached to simplices are algebraically special. Thus we do not have to worry about a continuum limit, since a finite result is exact.

Motivated by these considerations, I propose to consider the hypothesis:

Conjecture *The universe is in the CSW state.*

This idea is similar to the suggestion of Witten that the CSW theory is the topological phase of quantum gravity [2]. My suggestion differs from his in assuming that the universe is still in the topological phase, and that the concrete geometry we see is not the result of a fluctuation of the state of the universe as a whole, but rather due to the collapse of the wave packet in the presence of classical observers. As I shall explain below, the machinery of CSW theory provides spaces of states for parts of the universe, which I interpret as the relative states an observer might see. This suggestion is closely related to the many-worlds interpretation of quantum mechanics: the states on surfaces are data in one particular world. The state of the universe is the sum of all possible worlds, which takes the elegant form of the CSW invariant of links, or more concretely, of the Jones polynomial or one of its generalizations.

During the talk I gave at the conference one questioner pointed out, quite correctly, that it is not necessary to make this conjecture, since other states exist. Let me emphasize that it is only a hypothesis. I think it is fair to say that it leads to more beautiful mathematics than any other I know of. Whether that is a good guide for physics, only time will tell.

Topological quantum field theory in dimensions 2, 3, and 4 is leading into an extraordinarily rich mathematical field. In the spirit of the drunkard who looks for his keys near a street light, I have been thinking for several years about an attempt to unite the structures of TQFT with a suitable reinterpretation of quantum mechanics for the entire universe. I believe that the two subjects resemble one another strongly enough that the program I am outlining becomes natural.

2   Quantum mechanics of the universe. Categorical physics

I want to propose that a quantum field theory is the wrong structure to describe the entire universe. Quantum field theory, like quantum mechanics generally, presupposes an external classical observer. There can be no probability interpretation for the whole universe. In fact, the state of the universe cannot change, because there is no time, except in the presence of an external clock. Since the universe as a whole is in a fixed state, the quantum theory of the gravitational field as a whole is not a physical theory. It describes all possible universes, while the task of a physical theory is to explain measurements made in **this** universe in **this** state. Thus, what is needed is a theory which describes a universe in a particular state. This is similar to what some researchers, such as Penrose, have suggested: that the initial conditions of the universe are determined by the laws. What I am proposing is that the initial conditions become part of the laws. Naturally, this requires that the universe be in a very special state.

Philosophically, this is similar to the idea of a wave function of the universe. One can phrase this proposal as the suggestion that the wave function of the universe is the CSW functional. There is also a good deal here philosophically in common with the recent paper of Rovelli and Smolin [23]. Although they do not assume the universe is in the CSW state, they couple the gravitational field to a particular matter field, which acts as a clock, so that the gravitational field is treated as only part of a universe.

What replaces a quantum field theory is a family of quantum theories corresponding to parts of the universe. When we divide the universe into two parts, we obtain a Hilbert space which is associated to the boundary between them. This is another departure from quantum field theory; it amounts to abandonment of observation at a distance.

The quantum theories of different parts of the universe are not, of course, independent. In a situation where one observer watches another there must be maps from the space of states which one observer sees to the other. Furthermore, these maps must be consistent. If A watches B watch C watch the rest of the universe, A must see B see C see what A sees C see.

Since, in the Ashtekar/loop variables the states of quantum gravity are invariants of embedded graphs, we allow labelled punctures on the boundaries between parts of the universe, and include embedded graphs in the parts of the universe we study, i.e. in 3-manifolds with boundary.

The structure we arrive at has a natural categorical flavor. Objects are places where observations take place, i.e. boundaries; and morphisms are 3d cobordisms, which we think of as A observing B.

Let us formalize this as follows:

Definition 2.1   *An* **observer** *is an oriented 3-manifold with boundary* *containing an embedded labelled framed graph which intersects the boundary in isolated labelled points.*

(The labelling sets for the edges and vertices of the graph are finite, and need to be chosen for all of what ensues.)

Definition 2.2   *A **skin of observation** is a closed oriented surface with labelled punctures.*

Definition 2.3   *If A and B are skins of observation an **inspection**, α, of B **by** A is an observer whose boundary is identified with $\overline{A} \cup B$ (i.e. reverse the orientation on A) such that the labellings of the components of the graph which reach the boundary of α match the labellings in $\overline{A} \cup B$.*

This definition requires that the set of labellings possess an involution corresponding to reversal of orientation on a surface.

To every inspection α of B by A there corresponds a dual inspection of $\overline{A}$ by $\overline{B}$, given by reversing orientation and dualizing on α.

Definition 2.4   *The **category of observation** is the category whose objects are skins of observation and whose morphisms are inspections.*

Definition 2.5   *If M^3 is a closed oriented 3-manifold the **category of observation in** M^3 is the relative (i.e. embedded) version of the above.*

Of course, we can speak of observers, etc., in M^3.

Definition 2.6   *A **state for quantum gravity** (in M^3) is a functor from the category of observation (in M^3) to the category of vector spaces.*

Nowhere in any of this do we assume that these boundaries are connected. In fact, the most important situation to study may be one in which the universe is crammed full of a 'gas' of classical observers. We shall discuss this situation below as a key to a physical interpretation of our theory.

If we examine the mathematical structure necessary to produce a 'state' for the universe in this sense, we find that it is identical to a 3d TQFT. Thus, the CSW state produces a state in this new sense as well.

Another way to look at this proposal is as follows: the CSW state for the Ashtekar variables is a very special state, in that it factorizes when we cut the 3-manifold along a surface with punctures, so that a finite-dimensional Hilbert space is attached to each such surface, and the invariant of any link cut by the surface can be expressed as the inner product of two vectors corresponding to the two half links. (John Baez has pointed out that these finite-dimensional spaces really do possess natural inner products.) Thus, it produces a 'state' also in the sense we have defined above.

I interpret this as saying that the state of the universe is unchanging, but that because the universe is in a very special state, it can contain a classical world, i.e. a family of classical observers with consistent mutual observations. The states on pieces of the universe (i.e. links with ends in manifolds with boundary) can be interpreted as changing, once we learn to interpret vectors in the Hilbert spaces as clocks.

So far, we have a net of finite-dimensional Hilbert spaces, rather than one big one, and no idea how to reintroduce time in the presence of observers. There are natural things to try for both of these problems in the mathematical context of TQFT.

Before going into a program for solving these problems (and thereby opening the possibility of computing the results of real experiments) let us make a survey of the mathematical toolkit we inherit from TQFT.

3   Ideas from TQFT

As we have indicated, the notion of a 'state of quantum gravity' we defined above is equivalent to the notion of a 3d TQFT which is currently prevalent in the literature.

The simplest definition of a 3d TQFT is that it is a machine which assigns a vector space to a surface, and a linear map to a cobordism between two surfaces. The empty surface receives a 1-dimensional vector space, so a closed 3-manifold gets a numerical invariant. Composition of cobordisms corresponds to composition of the linear maps. A more abstract way to phrase this is that a TQFT is a functor from the cobordism category to the category of vector spaces.

The TQFTs which have appeared lately are richer than this. They assign vector spaces to surfaces with labelled punctures, and linear maps to cobordisms containing links or knotted graphs with labelled edges. The labels correspond to representations. Since it is easy to extend the loop variables to allow traces in arbitrary representations (characters), there is a great deal of coincidence in the pictures of CSW TQFT and the loop variables. (That constituted much of the initial motivation for this program.) We can describe this as a functor from a richer cobordism category to *Vect*. The objects in the richer cobordism category are surfaces with labelled punctures, and the morphisms are cobordisms containing labelled links with ends on the punctures.

There are several ways to construct TQFTs in various dimensions. Let us here discuss the construction of a TQFT from a triangulation. We assign labels to edges in the triangulation, and some sort of factors combining the edge labels to different dimensional simplices in the triangulation, multiply the factors together, and sum over labellings.

In order to obtain a topologically invariant theory, we need the combination factors to satisfy some equations. The equations they need to satisfy are very algebraic in nature; as we go through different classes of theories in different low dimensions we first rediscover most of the interesting classes of associative algebras, then of tensor categories [19].

FIG. 1. Move for 2d TQFT

A simple example in two dimensions may explain why the equations for a topological theory have a fundamental algebraic flavor. For a 2d TQFT defined from a triangulation, we need invariance under the move in Fig. 1.

Now, if we think of the coefficients which we use to combine the labels around a triangle as the structure coefficients of an algebra, this is exactly the associative law. Much of the rest of classical abstract algebra makes an appearance here too.

As has been described elsewhere [3], a 3d TQFT can be constructed from a modular tensor category. The interesting examples of MTCs can be realized as quotients of the categories of representations of quantum groups.

If we try to use the modular tensor categories to construct a TQFT on a triangulation, we obtain, not the CSW theory, but the weaker TQFT of Turaev and Viro [20]. Here we label edges, not from a basis for an algebra, but with irreducible objects from a tensor category. The combinations we attach to tetrahedra (3-simplices) are the quantum $6j$ symbols, which come from the associativity isomorphisms of the category.

We refer to this sort of formula as a state summation.

The full CSW theory can also be reproduced from a modular tensor category by a slightly subtler construction which uses a Heegaard splitting (which can be easily produced from a triangulation) [3].

The quantum $6j$ symbols satisfy some identities, which imply that the summation formula for a 3-manifold is independent under a change of the triangulation. The most interesting of these is the Elliot–Biedenharn identity. This identity is the consistency relation for the associativity isomorphism of the MTC. This is another example of the marriage between algebra and topology which underlies TQFT.

The suggestion that this sort of summation could be related to the quantum theory of gravity is older than one might think. If we use the representations of an ordinary Lie group instead of a quantum group, then the Turaev–Viro formula becomes the Ponzano–Regge formula for the evaluation of a spin network.

Ponzano and Regge [9], were able to interpret the evaluation as a sort of discrete path integral for 3d Euclidean quantum gravity. The formula which Regge and Ponzano found for the evaluation of a graph is the analog for a Lie group of the state sum of Turaev and Viro for a quantum group.

Thus, the evaluation of a tetrahedron for a spin network is a $6j$ symbol for the group $SU(2)$:

$$\sum_{labellings} \prod_{internal\ edges} \dim_q(j) \prod_{tetrahedra} \{6j\}. \tag{3.1}$$

The topological invariance of this formula follows from some elementary properties of representation theory. In (3.1), we have placed our trivalent labelled graph on the boundary of a 3-manifold with boundary, and then cut the interior up into tetrahedra, labelling the edges of all the internal lines with arbitrary spins. (The Clebsch–Gordan relations imply that only finitely many terms in (3.1) are nonzero.)

Ponzano and Regge then proceeded to interpret (3.1) as a discretized path integral for 3d quantum GR. The geometric interpretation consisted in thinking of the Casimirs of the representations as lengths of edges. Representation theory implied that the summation was dominated by flat geometries [9].

Another way to think of the program in this chapter is as an attempt to find the appropriate algebraic structure for extending the spin network approach to quantum gravity from d=3 to d=4.

Another idea from TQFT, which seems to have relevance for this physical program, is the ladder of dimensions [19]. TQFTs in adjacent dimensions seem to be related algebraically. The spirit of the relationship is like the relationship of a tensor category to an algebra. Tensor categories look just like algebras with the operation symbols in circles. The identities of an algebra correspond to isomorphisms in a tensor category. These isomorphisms then satisfy higher-order consistency or 'coherence' relations, which relate to topology up 1 dimension.

The program of construction in TQFT, which I hope will yield the tools for the quantization of general relativity, is not yet completed, but is progressing rapidly. There are two outstanding problems, which are mathematically natural, and which I believe are crucial to the physical problem as well. They are:

Topological problem 1 *Find a triangulation version of CSW theory (not merely its absolute square as in [20]).*

Topological problem 2 *Construct a 4d TQFT related via the dimensional ladder to CSW theory.*

The solution of these two problems will be in reach, if the program in [19] succeeds. The program of [19] also suggests that the solutions of these two topological problems are closely related, both being constructed from state sums involving the same new algebraic tools.

In this context we should also note that a 4d TQFT has been constructed in [21], out of a modular tensor category. This construction begins with a triangulation of a 4-manifold, and picks a Heegaard splitting for the boundary of each 4-simplex of the triangulation. Thus the theory in [21] can be thought of as producing a 4d theory by filling the 4-space with a network of 3d subspaces containing observers. This connection between TQFTs in 3 dimensions and 4 dimensions is the sort of thing I believe we will need to solve the problem of reintroducing time in a universe in a TQFT state. What I expect is that the program of [19] will provide richer examples of such a connection, which will prove to be relevant to quantizing general relativity.

What we conjecture in [19] is that a new algebraic structure will give us a new state summation, similar to the one we discussed above, which will give us CSW in 3 dimensions. If we use the new summation in 4 dimensions, we should obtain a discrete version of $F \wedge F$ theory. This combination seems very suggestive, since we are supposed to be obtaining our relative states from CSW theory on a boundary, while $F \wedge F$ theory is a Lagrangian for the topological sector for the quantum theory of GR [22]. Also, the algebraic structure we need to construct seems to be an expanded quantum version of the Lorentz group. It is this new, not yet understood summation, which comes from the representation theory of the new structure which we call an F algebra in [19], which I believe should give us a discretized version of a path integral for quantum gravity.

4   Hilbert space is dead—long live Hilbert space *or* the observer gas approximation

Let us assume that we have found a suitable state summation formula to solve our two topological problems (so that what we suggest will be predicated on the success of the program in [19], although one could also try to use the state sum in [21] together with Turaev–Viro theory). There is a natural proposal to make a physical interpretation within it in a 4-dimensional setting. In the scheme I am suggesting, we can recover a Hilbert space in a certain sort of classical limit, in which the universe is full of a 'gas', of classical observers. The idea is that if we choose a triangulation for the 4-manifold and a Heegaard splitting for the boundary of each 4-simplex, then we obtain a family of 2-surfaces which can be thought of as filling up the 4-manifold. These surfaces should be thought of as skins of observation for a family of observers who are crowding the spacetime. Now let us think of a 4-manifold with boundary. In a 3-dimensional boundary component, we can then combine the vector spaces which we assign to the surfaces which cut the boundaries of those 4-simplices lying on the boundary into a larger Hilbert space, which is a quotient of their tensor product. (It is a quotient because the surfaces overlap.) Now, if we pass to a finer triangulation, we will obtain a larger space, with a linear map to the smaller one. The linear map comes from the fact that we are using a 3d TQFT, and the existence of 3d cobordisms connecting pieces of the surfaces of the two sets of Heegaard splittings.

We end up with a large vector space for each triangulation of the boundary 3-manifold, and a linear map when one triangulation is a refinement of another. This produces a directed graph of vector spaces and linear maps associated to a 3-manifold. The 4d state sum can be extended to act on the vectors in these vector spaces by extending a triangulation for the boundary 3-manifold to one for the 4-manifold.

Whenever we have a directed set of vector spaces and consistent linear maps, there is a construction called the inverse limit which can combine them into a single vector space. The vectors are vectors in any of the spaces, with images under the maps identified. Since the 4d state sum is assumed to be topologically invariant, we can extend the linear map it assigns to a 4d cobordism to a map on the inverse limit spaces. I propose that it is these inverse limit spaces which play the role of the physical Hilbert spaces of the theory.

The thought here is that physical Hilbert space is the union of the finite-dimensional spaces which a gas of observers can see, in the limit of an infinitely dense gas.

The effect of this suggestion is to reverse the relationship between the finite-dimensional spaces of states which each observer can actually see and the global Hilbert space. We are regarding the states which can be observed at one skin of observation as primary. This 'relational' approach bears some resemblance to the physical ideas of Leibniz, although that is not an argument for it.

5 The problem of time

The test of this proposal is whether it can reproduce Einstein's equations in some classical limit. What I am proposing is that the 4d state sum which I hope to construct from the representation theory of the F algebra can be interpreted as a discretized version of the path integral for general relativity. The key question is whether the state sum is dominated in the limit of large spins on edges by assignments of spins whose geometric interpretation would give a discrete approximation to an Einstein manifold. This would mean that the theory was a quantum theory whose classical limit was general relativity.

The analogy with the work of Ponzano and Regge is the first suggestion that this program might work. In [9], they interpreted the summation we wrote down above as a discretized path integral for 3d quantum gravity. The geometric interpretation consisted in thinking of the Casimirs of representations as lengths. It was somewhat easier to get Einstein's equations in $d = 3$, since the solutions are just flat metrics.

Should this program work? One objection, which was raised in the discussion, is that the $F \wedge F$ model may only be a sector of quantum gravity in some weak sense. (I am restating this objection somewhat, in the absence of the questioner.) It can be replied that since the theory we are writing has the right symmetries, the action of the renormalization group should fix the Lagrangian. I do not think that either the objection or the reply is decisive in the abstract. If the state sum suggested in [19] can be defined, then we can investigate whether we recover Einstein's equation or not.

If this does work, one could do the state sum on a 4-manifold with corners, i.e. fix states on surfaces in the 3-dimensional boundaries of a 4d cobordism. One could interpret these states as initial conditions for an experiment, and investigate the results by studying the propagation via the 4d state sum.

One could conjecture that this would give a spacetime picture for the evolution of relative states within the framework of the CSW state of the universe.

6 Matter and symmetry

It is natural to ask whether this picture could be expanded to include matter fields. The obvious thing to try is to pass from $SU(2)$ to a larger group. It may be noted that Peldan [12] has observed that the CSW state also satisfies the Yang–Mills equation. If we really get a picture of gravity from the state sum associated to $SU(2)$, it would not be a great leap to try a larger gauge group.

As I was writing this I became aware of [24], in which Gambini and Pullin point out that the CSW state is also a state for GR coupled to electromagnetism. They also say that the sources for such a theory, thought of as places where links have ends, are necessarily fermionic. These results seem to strengthen further the program set forth here.

In general, the constructions which lead up to the F algebras in [19] can be thought of as expressions of 'quantum symmetry', i.e. as relatives of Lie groups in a quantum context. For example, the modular tensor categories can be described as deformed Clebsch–Gordan coefficients. The F algebras arise as a result of pushing this process further, recategorifying the categories into 2-categories. These constructions can also be done from extremely simple starting points, looking at representations of Dynkin diagrams as 'quivers'. The idea that theories in physics should be reconstructed from symmetries actually originated with Einstein, who was thinking of general relativity. It would be fitting somehow to reconstruct truly fundamental theories of physics from fundamental mathematical expressions of symmetries.

Acknowledgements

The author wishes to thank Lee Smolin and Carlo Rovelli for many years of conversation on this extremely elusive topic. The mathematical ideas here are largely the result of collaborations with David Yetter and Igor Frenkel. Louis Kauffman has provided many crucial insights. The author is supported by NSF grant DMS-9106476.

Bibliography

1. C. Rovelli and L. Smolin, Loop representation for quantum general relativity, *Nucl. Phys.* **B331** (1990), 80–152.

2. E. Witten, Quantum field theory and the Jones polynomial, *Commun. Math. Phys.* **121** (1989), 351–399.

3. L. Crane, 2-d physics and 3-d topology, *Commun. Math. Phys.* **135** (1991), 615–640.

4. G. Moore and N. Seiberg, Classical and quantum conformal field theory, *Commun. Math. Phys.* **123** (1989), 177–254.

5. B. Brügmann, R. Gambini, and J. Pullin, Jones polynomials for intersecting knots as physical states for quantum gravity, *Nucl. Phys.* **B385** (1992), 587–603.

6. L. Crane, Quantum symmetry, link invariants and quantum geometry, in *Proceedings of the XXth International Conference on Differential Geometric Methods in Theoretical Physics*, eds S. Catto and A. Rocha, Singapore, World Scientific, 1992.

7. L. Crane, Conformal field theory, spin geometry, and quantum gravity, *Phys. Lett.* **B259** (1991), 243–248.

8. R. Penrose, Angular momentum; an approach to combinatorial space time, in *Quantum Theory and Beyond,* ed. T. Bastin, Cambridge University Press, 1971.

9. G. Ponzano and T. Regge, Semiclassical limits of Racah coefficients, in *Spectroscopic and Group Theoretical Methods in Physics,* ed. F. Bloch, New York, Wiley, 1968.

10. C. Rovelli, What is an observable in classical and quantum gravity? *Classical & Quantum Gravity* **8** (1991), 297–316.

11. J. Moussouris, Quantum Models of Space Time Based on Recoupling Theory, Thesis, Oxford University (1983).

12. P. Peldan, Ashtekar's variables for arbitrary gauge group, *Phys. Rev.* **D46** (1992), R2279–R2282.

13. L. Smolin, personal communication.

14. V. Moncrief, personal communication.

15. K. Walker, On Witten's 3-manifold invariants, unpublished.

16. L. Crane and I. Frenkel, Hopf categories and their representations,

to appear.

17. M. Kapranov and V. Voevodsky, Braided monoidal 2-categories, 2-vector spaces and Zamolodchikov's tetrahedra equation, preprint.

18. C. Rovelli and L. Smolin, Knot theory and quantum gravity, *Phys. Rev. Lett.* **61** (1988), 1155–1158.

19. L. Crane and I. Frenkel, A representation theoretic approach to 4-dimensional topological quantum field theory, to appear in Proceedings of AMS conference, Mt Holyoke, Massachusetts, June 1992.

20. V. Turaev and O. Viro, State sum invariants of 3-manifolds and quantum $6j$ symbols, *Topology* **31** (1992), 865–902.

21. L. Crane and D. Yetter, A categorical construction of 4d TQFT's, to appear in Proceedings of AMS conference, Dayton, Ohio, October 1992.

22. L. N. Chang and C. Soo, BRST cohomology and invariants of 4D gravity in Ashtekar variables, preprint.

23. C. Rovelli and L. Smolin, The physical hamiltonian in non-perturbative quantum gravity, preprint.

24. R. Gambini and J. Pullin, Quantum Einstein–Maxwell fields: a unified viewpoint from the loop representation, preprint.

Strings, Loops, Knots, and Gauge Fields

John C. Baez

Department of Mathematics, University of California,
Riverside, California 92521, USA
(email: jbaez@ucrmath.ucr.edu)

Abstract

The loop representation of quantum gravity has many formal resemblances to a background-free string theory. In fact, its origins lie in attempts to treat the string theory of hadrons as an approximation to QCD, in which the strings represent flux tubes of the gauge field. A heuristic path-integral approach indicates a duality between background-free string theories and generally covariant gauge theories, with the loop transform relating the two. We review progress towards making this duality rigorous in three examples: 2d Yang–Mills theory (which, while not generally covariant, has symmetry under all area-preserving transformations), 3d quantum gravity, and 4d quantum gravity. $SU(N)$ Yang–Mills theory in 2 dimensions has been given a string-theoretic interpretation in the large-N limit, but here we provide an exact string-theoretic interpretation of the theory on $\mathbf{R} \times S^1$ for finite N. The string-theoretic interpretation of quantum gravity in 3 dimensions gives rise to conjectures about integrals on the moduli space of flat connections, while in 4 dimensions there may be relationships to the theory of 2-tangles.

1   Introduction

The notion of a deep relationship between string theories and gauge theories is far from new. String theory first arose as a model of hadron interactions. Unfortunately this theory had a number of undesirable features; in particular, it predicted massless spin-2 particles. It was soon supplanted by quantum chromodynamics (QCD), which models the strong force by an $SU(3)$ Yang–Mills field. However, string models continued to be popular as an approximation of the confining phase of QCD. Two quarks in a meson, for example, can be thought of as connected by a string-like flux tube in which the gauge field is concentrated, while an excitation of the gauge field alone can be thought of as a looped flux tube. This is essentially a modern reincarnation of Faraday's notion of 'field lines', but it can be formalized using the notion of Wilson loops. If A denotes the vector potential, or connection, in a classical gauge theory, a Wilson loop is simply the trace of the holonomy of A around a loop γ in space, typically written in terms of a path-ordered exponential

$$\mathrm{Tr}\, \mathrm{P}\, e^{\oint_\gamma A}.$$

If instead A denotes the vector potential in a quantized gauge theory, the Wilson loop may be reinterpreted as an operator on the Hilbert space of states, and, at least heuristically, applying this operator to the vacuum state one obtains a state in which the Yang–Mills analog of the electric field flows around the loop γ.

In the late 1970s, Makeenko and Migdal, Nambu, Polyakov, and others [40, 45] attempted to derive equations of string dynamics as an approximation to the Yang–Mills equation, using Wilson loops. More recently, D. Gross and others [26, 27, 37, 38, 39] have been able to reformulate Yang–Mills theory in 2-dimensional spacetime as a string theory by writing an asymptotic series for the vacuum expectation values of Wilson loops as a sum over maps from surfaces (the string worldsheet) to spacetime. This development raises the hope that other gauge theories might also be isomorphic to string theories. For example, recent work by Witten [56] and Periwal [44] suggests that Chern–Simons theory in 3 dimensions is also equivalent to a string theory.

String theory eventually became popular as a theory of everything because the massless spin-2 particles it predicted could be interpreted as the gravitons one obtains by quantizing the spacetime metric perturbatively about a fixed 'background' metric. Since string theory appears to avoid the renormalization problems in perturbative quantum gravity, it is a strong candidate for a theory unifying gravity with the other forces. However, while classical general relativity is an elegant geometrical theory relying on no background structure for its formulation, it has proved difficult to describe string theory along these lines. Typically one begins with a fixed background structure and writes down a string field theory in terms of this; only afterwards can one investigate its background independence [58]. The clarity of a manifestly background-free approach to string theory would be highly desirable.

On the other hand, attempts to formulate Yang–Mills theory in terms of Wilson loops eventually led to a full-fledged 'loop representation' of gauge theories, thanks to the work of Gambini and Trias [22], and others. After Ashtekar [1] formulated quantum gravity as a sort of gauge theory using the 'new variables', Rovelli and Smolin [49] were able to use the loop representation to study quantum gravity non-perturbatively in a manifestly background-free formalism. While superficially quite different from modern string theory, this approach to quantum gravity has many points of similarity, thanks to its common origin. In particular, it uses the device of Wilson loops to construct a space of states consisting of 'multiloop invariants', which assign an amplitude to any collection of loops in space.

The resemblance of these states to wavefunctions of a string field theory is striking. It is natural, therefore, to ask whether the loop representation of quantum gravity might be a string theory in disguise—or vice versa.

The present paper does not attempt a definitive answer to this question. Rather, we begin by describing a general framework relating gauge theories and string theories, and then consider a variety of examples. Our treatment of examples is also meant to serve as a review of Yang–Mills theory in 2 dimensions and quantum gravity in 3 and 4 dimensions.

In Section 2 we describe how the loop representation of a generally covariant gauge theory is related to a background-free closed string field theory. We take a very naive approach to strings, thinking of them simply as maps from a surface into spacetime, and disregarding any conformal structure or fields propagating on the surface. We base our treatment on the path integral formalism, and in order to simplify the presentation we make a number of overoptimistic assumptions concerning measures on infinite-dimensional spaces such as the space $\mathcal{A}/\mathcal{G}$ of connections modulo gauge transformations.

In Section 3 we consider Yang-Mills theory in 2 dimensions as an example. In fact, this theory is not generally covariant, but it has an infinite-dimensional subgroup of the diffeomorphism group as symmetries, the group of all area-preserving transformations. Rather than the path integral approach we use canonical quantization, which is easier to make rigorous. Gross, Taylor, and others [17, 26, 27, 37, 38] have already given 2-dimensional $SU(N)$ Yang–Mills theory a string-theoretic interpretation in the large-N limit. Our treatment is mostly a review of their work, but we find it to be little extra effort, and rather enlightening, to give the theory a precise string-theoretic interpretation for finite N.

In Section 4 we consider quantum gravity in 3 dimensions. We review the loop representation of this theory and raise some questions about integrals over the moduli space of flat connections on a Riemann surface whose resolution would be desirable for developing a string-theoretic picture of the theory. We also briefly discuss Chern–Simons theory in 3 dimensions.

These examples have finite-dimensional reduced configuration spaces, so there are no analytical difficulties with measures on infinite-dimensional spaces, at least in canonical quantization. In Section 5, however, we consider quantum gravity in 4 dimensions. Here the classical configuration space is infinite dimensional and issues of analysis become more important. We review recent work by Ashtekar, Isham, Lewandowski, and the author [3, 4, 10] on diffeomorphism-invariant generalized measures on $\mathcal{A}/\mathcal{G}$ and their relation to multiloop invariants and knot theory. We also note how a string-theoretic interpretation of the theory leads naturally to the study of 2-tangles.

2 String field/gauge field duality

In this section we sketch a relationship between string field theories and gauge theories. We begin with a non-perturbative Lagrangian description of background-free closed string field theories. From this we derive a Hamiltonian description, which turns out to be mathematically isomorphic to the loop representation of a generally covariant gauge theory. We emphasize that while our discussion here is rigorous, it is schematic, in the sense that some of our assumptions are not likely to hold precisely as stated in the most interesting examples. In particular, by 'measure' in this section we will always mean a positive regular Borel measure, but in fact one should work with a more general version of this concept. We discuss these analytical issues more carefully in Section 5.

Consider a theory of strings propagating on a spacetime M that is diffeomorphic to $\mathbf{R} \times X$, with X a manifold we call 'space'. We do not assume a *canonical* identification of M with $\mathbf{R} \times X$, or any other background structure (metric, etc.) on spacetime. We take the classical configuration space of the string theory to be the space $\mathcal{M}$ of multiloops in X:

$$\mathcal{M} = \bigcup_{n \geq 0} \mathcal{M}_n$$

with

$$\mathcal{M}_n = \mathrm{Maps}(nS^1, X).$$

Here nS^1 denotes the disjoint union of n copies of S^1, and we write 'Maps' to denote the set of maps satisfying some regularity conditions (continuity, smoothness, etc.) to be specified. Let $\mathcal{D}\gamma$ denote a measure on $\mathcal{M}$ and let $\mathrm{Fun}(\mathcal{M})$ denote some space of square-integrable functions on $\mathcal{M}$. We assume that $\mathrm{Fun}(\mathcal{M})$ and the measure $\mathcal{D}\gamma$ are invariant both under diffeomorphisms of space and reparametrizations of the strings. That is, both the identity component of the diffeomorphism group of X and the orientation-preserving diffeomorphisms of nS^1 act on $\mathcal{M}$, and we wish $\mathrm{Fun}(\mathcal{M})$ and $\mathcal{D}\gamma$ to be preserved by these actions.

Introduce on $\mathrm{Fun}(M)$ the 'kinematical inner product', which is just the L^2 inner product

$$\langle \psi, \phi \rangle_{kin} = \int_{\mathcal{M}} \overline{\psi}(\gamma) \phi(\gamma) \, \mathcal{D}\gamma.$$

We assume for convenience that this really is an inner product, i.e. it is non-degenerate. Define the 'kinematical state space' $\mathbf{H}_{kin}$ to be the Hilbert space completion of $\mathrm{Fun}(\mathcal{M})$ in the norm associated to this inner product.

Following ideas from canonical quantum gravity, we do not expect $\mathbf{H}_{kin}$ to be the true space of physical states. In the space of physical states, any two states differing by a diffeomorphism of spacetime are identified. The physical state space thus depends on the dynamics of the theory. Taking a

Lagrangian approach, dynamics may be described in terms of path integrals as follows. Fix a time $t > 0$. Let $\mathcal{P}$ denote the set of 'histories', that is maps $f: \Sigma \to [0, t] \times X$, where Σ is a compact oriented 2-manifold with boundary, such that

$$f(\Sigma) \cap \partial([0, t] \times X) = f(\partial \Sigma).$$

Given $\gamma, \gamma' \in \mathcal{M}$, we say that $f \in \mathcal{P}$ is a history from γ to γ' if $f: \Sigma \to [0, t] \times X$ and the boundary of Σ is a disjoint union of circles $nS^1 \cup mS^1$, with

$$f|nS^1 = \gamma, \qquad f|mS^1 = \gamma'.$$

We fix a measure, or 'path integral', on $\mathcal{P}(\gamma, \gamma')$. Following tradition, we write this as $e^{iS(f)} \mathcal{D}f$, with $S(f)$ denoting the action of f, but $e^{iS(f)}$ and $\mathcal{D}f$ only appear in the combination $e^{iS(f)} \mathcal{D}f$. Since we are interested in generally covariant theories, this path integral is assumed to have some invariance properties, which we note below as they are needed.

Using the standard recipe in topological quantum field theory, we define the 'physical inner product' on $\mathbf{H}_{kin}$ by

$$\langle \psi, \phi \rangle_{phys} = \int_{\mathcal{M}} \int_{\mathcal{M}} \int_{\mathcal{P}(\gamma, \gamma')} \overline{\psi}(\gamma) \phi(\gamma') \, e^{iS(f)} \mathcal{D}f \mathcal{D}\gamma \mathcal{D}\gamma'$$

assuming optimistically that this integral is well defined. We do not actually assume this is an inner product in the standard sense, for while we assume $\langle \psi, \psi \rangle \geq 0$ for all $\psi \in \mathbf{H}_{kin}$, we do not assume positive definiteness. The general covariance of the theory should imply that this inner product is independent of the choice of time $t > 0$, so we assume this as well.

Define the space of norm-zero states $\mathbf{I} \subseteq \mathbf{H}_{kin}$ by

$$\mathbf{I} = \{\psi | \langle \psi, \psi \rangle_{phys} = 0\}$$
$$= \{\psi | \langle \psi, \phi \rangle_{phys} = 0 \text{ for all } \phi \in \mathbf{H}_{kin}\}$$

and define the 'physical state space' $\mathbf{H}_{phys}$ to be the Hilbert space completion of $\mathbf{H}_{kin}/\mathbf{I}$ in the norm associated to the physical inner product. In general $\mathbf{I}$ is non-empty, because if $g \in \text{Diff}_0(X)$ is a diffeomorphism in the connected component of the identity, we can find a path of diffeomorphisms $g_t \in \text{Diff}_0(M)$ with $g_0 = g$ and g_t equal to the identity, and defining $\widetilde{g} \in \text{Diff}([0, t] \times X)$ by

$$\widetilde{g}(t, x) = (t, g_t(x)),$$

we have

$$\langle \psi, \phi \rangle_{phys} = \int_{\mathcal{M}} \int_{\mathcal{M}} \int_{\mathcal{P}(\gamma, \gamma')} \overline{\psi}(\gamma) \phi(\gamma') e^{iS(f)} \, \mathcal{D}f \mathcal{D}\gamma \mathcal{D}\gamma'$$
$$= \int_{\mathcal{M}} \int_{\mathcal{M}} \int_{\mathcal{P}(\gamma, \gamma')} \overline{g\psi}(g\gamma) \phi(\gamma') \, e^{iS(f)} \mathcal{D}f \mathcal{D}\gamma \mathcal{D}\gamma'$$

$$= \int_{\mathcal{M}} \int_{\mathcal{M}} \int_{\mathcal{P}(\gamma,\gamma')} \overline{g\psi}(g\gamma)\phi(\gamma')\, e^{iS(\tilde{g}f)}\, \mathcal{D}(\tilde{g}f)\mathcal{D}(g\gamma)\mathcal{D}\gamma'$$

$$= \int_{\mathcal{M}} \int_{\mathcal{M}} \int_{\mathcal{P}(\gamma,\gamma')} \overline{g\psi}(\gamma)\phi(\gamma')\, e^{iS(f)}\mathcal{D}f\mathcal{D}\gamma\mathcal{D}\gamma'$$

$$= \langle g\psi, \phi \rangle_{phys}$$

for any ψ, ϕ. Here we are assuming

$$e^{iS(\tilde{g}f)}\, \mathcal{D}(\tilde{g}f) = e^{iS(f)}\mathcal{D}f,$$

which is one of the expected invariance properties of the path integral. It follows that $\mathbf{I}$ includes the space $\mathbf{J}$, the closure of the span of all vectors of the form $\psi - g\psi$. We can therefore define the (spatially) 'diffeomorphism-invariant state space' $\mathbf{H}_{diff}$ by $\mathbf{H}_{diff} = \mathbf{H}_{kin}/\mathbf{J}$ and obtain $\mathbf{H}_{phys}$ as a Hilbert space completion of $\mathbf{H}_{diff}/\mathbf{K}$, where $\mathbf{K}$ is the image of $\mathbf{I}$ in $\mathbf{H}_{diff}$.

To summarize, we obtain the physical state space from the kinematical state space by taking two quotients:

$$\mathbf{H}_{kin} \to \mathbf{H}_{kin}/\mathbf{J} = \mathbf{H}_{diff}$$
$$\mathbf{H}_{diff} \to \mathbf{H}_{diff}/\mathbf{K} \hookrightarrow \mathbf{H}_{phys}.$$

As usual in canonical quantum gravity and topological quantum field theory, there is no Hamiltonian; instead, all the information about dynamics is contained in the physical inner product. The reason, of course, is that the path integral, which in traditional quantum field theory described time evolution, now describes the physical inner product. The quotient map $\mathbf{H}_{diff} \to \mathbf{H}_{phys}$, or equivalently its kernel $\mathbf{K}$, plays the role of a 'Hamiltonian constraint'. The quotient map $\mathbf{H}_{kin} \to \mathbf{H}_{diff}$, or equivalently its kernel $\mathbf{J}$, plays the role of the 'diffeomorphism constraint', which is independent of the dynamics. (Strictly speaking, we should call $\mathbf{K}$ the 'dynamical constraint', as we shall see that it expresses constraints on the initial data other than those usually called the Hamiltonian constraint, such as the 'Mandelstam constraints' arising in gauge theory.)

It is common in canonical quantum gravity to proceed in a slightly different manner than we have done here, using subspaces at certain points where we use quotient spaces [49, 50]. For example, $\mathbf{H}_{diff}$ may be defined as the subspace of $\mathbf{H}_{kin}$ consisting of states invariant under the action of $\text{Diff}_0(X)$, and $\mathbf{H}_{phys}$ then defined as the kernel of certain operators, the Hamiltonian constraints. The method of working solely with quotient spaces has, however, been studied by Ashtekar [2].

The choice between these different approaches will in the end be dictated by the desire for convenience and/or rigor. As a heuristic guiding principle, however, it is worth noting that the subspace and quotient space approaches are essentially equivalent if we assume that the subspace $\mathbf{I}$ is closed in the norm topology on $\mathbf{H}_{kin}$. Relative to the kinematical inner

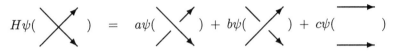

$$H\psi(\;\times\;) \;=\; a\psi(\;\times\;) \;+\; b\psi(\;\times\;) \;+\; c\psi(\;\equiv\;)$$

Fig. 1. Two-string interaction in 3-dimensional space

product, we can identify $\mathbf{H}_{diff}$ with the orthogonal complement $\mathbf{J}^{\perp}$, and similarly identify $\mathbf{H}_{phys}$ with $\mathbf{I}^{\perp}$. From this point of view we have

$$\mathbf{H}_{phys} \subseteq \mathbf{H}_{diff} \subseteq \mathbf{H}_{kin}.$$

Moreover, $\psi \in \mathbf{H}_{diff}$ if and only if ψ is invariant under the action of $\mathrm{Diff}_0(X)$ on $\mathbf{H}_{kin}$. To see this, first note that if $g\psi = \psi$ for all $g \in \mathrm{Diff}_0(X)$, then for all $\phi \in \mathbf{H}_{kin}$ we have

$$\langle \psi, g\phi - \phi \rangle = \langle g^{-1}\psi - \psi, \phi \rangle = 0$$

so $\psi \in \mathbf{J}^{\perp}$. Conversely, if $\psi \in \mathbf{J}^{\perp}$,

$$\langle \psi, g\psi - \psi \rangle = 0$$

so $\langle \psi, \psi \rangle = \langle \psi, g\psi \rangle$, and since g acts unitarily on $\mathbf{H}_{kin}$ the Cauchy–Schwarz inequality implies $g\psi = \psi$.

The approach using subspaces is the one with the clearest connection to knot theory. An element $\psi \in \mathbf{H}_{kin}$ is a function on the space of multiloops. If ψ is invariant under the action of $\mathrm{Diff}_0(X)$, we call ψ a 'multiloop invariant'. In particular, ψ defines an ambient isotopy invariant of links in X when we restrict it to links (which are nothing but multiloops that happen to be embeddings). We see therefore that in this situation the physical states define link invariants. As a suggestive example, take $X = S^3$, and take as the Hamiltonian constraint an operator H on $\mathbf{H}_{diff}$ that has the property described in Fig. 1. Here $a, b, c \in \mathbf{C}$ are arbitrary. This Hamiltonian constraint represents the simplest sort of diffeomorphism-invariant two-string interaction in 3-dimensional space. Defining the physical space $\mathbf{H}_{phys}$ to be the kernel of H, it follows that any $\psi \in \mathbf{H}_{phys}$ gives a link invariant that is just a multiple of the HOMFLY invariant [21]. For appropriate values of the parameters a, b, c, we expect this sort of Hamiltonian constraint to occur in a generally covariant gauge theory on 4-dimensional spacetime known as *BF* theory, with gauge group $SU(N)$ [29]. A similar construction working with unoriented framed multiloops gives rise to the Kauffman polynomial, which is associated with *BF* theory with gauge group $SO(N)$ [31]. We see here in its barest form the path from string-theoretic considerations to link invariants and then to gauge theory.

In what follows, we start from the other end, and consider a generally covariant gauge theory on M. Thus we fix a Lie group G and a principal G-bundle $P \to M$. Fixing an identification $M \cong \mathbf{R} \times X$, the classical configuration space is the space $\mathcal{A}$ of connections on $P|_{\{0\} \times X}$. (The physical

Hilbert space of the quantum theory, it should be emphasized, is supposed to be independent of this identification $M \cong \mathbf{R} \times X$.) Given a loop $\gamma \colon S^1 \to X$ and a connection $A \in \mathcal{A}$, let $T(\gamma, A)$ be the corresponding Wilson loop, that is the trace of the holonomy of A around γ in a fixed finite-dimensional representation of G:

$$T(\gamma, A) = \operatorname{Tr} \mathrm{P}\, e^{\oint_\gamma A}.$$

The group $\mathcal{G}$ of gauge transformations acts on $\mathcal{A}$. Fix a $\mathcal{G}$-invariant measure $\mathcal{D}A$ on $\mathcal{A}$ and let $\operatorname{Fun}(\mathcal{A}/\mathcal{G})$ denote a space of gauge-invariant functions on $\mathcal{A}$ containing the algebra of functions generated by Wilson loops. We may alternatively think of $\operatorname{Fun}(\mathcal{A}/\mathcal{G})$ as a space of functions on $\mathcal{A}/\mathcal{G}$ and $\mathcal{D}A$ as a measure on $\mathcal{A}/\mathcal{G}$. Assume that $\mathcal{D}A$ is invariant under the action of $\operatorname{Diff}_0(M)$ on $\mathcal{A}/\mathcal{G}$, and define the kinematical state space $\mathbf{H}_{kin}$ to be the Hilbert space completion of $\operatorname{Fun}(\mathcal{A}/\mathcal{G})$ in the norm associated to the kinematical inner product

$$\langle \psi, \phi \rangle_{kin} = \int_{\mathcal{A}/\mathcal{G}} \overline{\psi}(A)\phi(A)\, \mathcal{D}A.$$

The relation of this kinematical state space and that described above for a string field theory is given by the loop transform. Given any multiloop $(\gamma_1, \ldots, \gamma_n) \in \mathcal{M}_n$, define the loop transform $\hat{\psi}$ of $\psi \in \operatorname{Fun}(\mathcal{A}/\mathcal{G})$ by

$$\hat{\psi}(\gamma_1, \ldots, \gamma_n) = \int_{\mathcal{A}/\mathcal{G}} \psi(A)T(\gamma_1, A) \cdots T(\gamma_n, A)\, \mathcal{D}A.$$

Take $\operatorname{Fun}(\mathcal{M})$ to be the space of functions in the range of the loop transform. Let us assume, purely for simplicity of exposition, that the loop transform is one-to-one. Then we may identify $\mathbf{H}_{kin}$ with $\operatorname{Fun}(\mathcal{M})$ just as in the string field theory case.

The process of passing from the kinematical state space to the diffeomorphism-invariant state space and then the physical state space has already been treated for a number of generally covariant gauge theories, most notably quantum gravity [1, 48, 49]. In order to emphasize the resemblance to the string field case, we will use a path integral approach.

Fix a time $t > 0$. Given $A, A' \in \mathcal{A}$, let $\mathcal{P}(A, A')$ denote the space of connections on $P|_{[0,t] \times X}$ which restrict to A on $\{0\} \times X$ and to A' on $\{t\} \times X$. We assume the existence of a measure on $\mathcal{P}(A, A')$ which we write as $e^{iS(a)}\mathcal{D}a$, using a to denote a connection on $P|_{[0,t] \times X}$. Again, this generalized measure has some invariance properties corresponding to the general covariance of the gauge theory. Define the 'physical' inner product on $\mathbf{H}_{kin}$ by

$$\langle \psi, \phi \rangle_{phys} = \int_{\mathcal{A}} \int_{\mathcal{A}} \int_{\mathcal{P}(A,A')} \overline{\psi}(A)\phi(A')\, e^{iS(a)}\mathcal{D}a\mathcal{D}A\mathcal{D}A'$$

again assuming that this integral is well defined and that $\langle \psi, \psi \rangle \geq 0$ for all ψ. This inner product should be independent of the choice of time $t > 0$. Letting $\mathbf{I} \subseteq \mathbf{H}_{kin}$ denote the space of norm-zero states, the physical state space $\mathbf{H}_{phys}$ of the gauge theory is $\mathbf{H}_{kin}/\mathbf{I}$. As before, we can use the general covariance of the theory to show that $\mathbf{I}$ contains the closed span $\mathbf{J}$ of all vectors of the form $\psi - g\psi$. Letting $\mathbf{H}_{diff} = \mathbf{H}_{kin}/\mathbf{J}$, and letting $\mathbf{K}$ be the image of $\mathbf{I}$ in $\mathbf{H}_{diff}$, we again see that the physical state space is obtained by applying first the diffeomorphism constraint

$$\mathbf{H}_{kin} \to \mathbf{H}_{kin}/\mathbf{J} = \mathbf{H}_{diff}$$

and then the Hamiltonian constraint

$$\mathbf{H}_{diff} \to \mathbf{H}_{diff}/\mathbf{K} \hookrightarrow \mathbf{H}_{phys}.$$

In summary, we see that the Hilbert spaces for generally covariant string theories and generally covariant gauge theories have a similar form, with the loop transform relating the gauge theory picture to the string theory picture. The key point, again, is that a state ψ in $\mathbf{H}_{kin}$ can be regarded either as a wave function on the classical configuration space $\mathcal{A}$ for gauge fields, with $\psi(A)$ being the amplitude of a specified connection A, *or* as a wave function on the classical configuration space $\mathcal{M}$ for strings, with $\hat{\psi}(\gamma_1, \ldots, \gamma_n)$ being the amplitude of a specified n-tuple of strings $\gamma_1, \ldots, \gamma_n \colon S^1 \to X$ to be present. The loop transform depends on the non-linear 'duality' between connections and loops,

$$\mathcal{A}/\mathcal{G} \times \mathcal{M} \to \mathbf{C}$$
$$(A, (\gamma_1, \ldots, \gamma_n)) \mapsto T(A, \gamma) \cdots T(A, \gamma_n)$$

which is why we speak of string field/gauge field duality rather than an isomorphism between string fields and gauge fields.

At this point it is natural to ask what the difference is, apart from words, between the loop representation of a generally covariant gauge theory and the sort of purely topological string field theory we have been considering. From the Hamiltonian viewpoint (that is, in terms of the spaces $\mathbf{H}_{kin}$, $\mathbf{H}_{diff}$, and $\mathbf{H}_{phys}$) the difference is not so great. The Lagrangian for a gauge theory, on the other hand, is quite a different object than that of a string field theory. Note that nothing we have done allows the direct construction of a string field Lagrangian from a gauge field Lagrangian or vice versa. In the following sections we will consider some examples: Yang–Mills theory in 2 dimensions, quantum gravity in 3 dimensions, and quantum gravity in 4 dimensions. In no case is a string field action $S(f)$ known that corresponds to the gauge theory in question! However, in 2d Yang–Mills theory a working substitute for the string field path integral is known: a discrete sum over certain equivalence classes of maps $f \colon \Sigma \to M$. This is, in fact, a promising alternative to dealing with measures on the space $\mathcal{P}$ of string histories. In 4-dimensional quantum gravity, such an approach might involve a sum over '2-tangles', that is ambient isotopy classes of embeddings $f: \Sigma \to [0, t] \times X$.

3   Yang–Mills theory in 2 dimensions

We begin with an example in which most of the details have been worked out. Yang–Mills theory is not a generally covariant theory since it relies for its formulation on a fixed Riemannian or Lorentzian metric on the spacetime manifold M. We fix a connected compact Lie group G and a principal G-bundle $P \to M$. Classically the gauge fields in question are connections A on P, and the Yang–Mills action is given by

$$S(A) = -\frac{1}{2} \int_M \text{Tr}(F \wedge \star F)$$

where F is the curvature of A and Tr is the trace in a fixed faithful unitary representation of G and hence its Lie algebra $\mathbf{g}$. Extremizing this action we obtain the classical equations of motion, the Yang–Mills equation

$$d_A \star F = 0,$$

where d_A is the exterior covariant derivative.

The action $S(A)$ is gauge invariant, so it can be regarded as a function on the space of connections on M modulo gauge transformations. The group $\text{Diff}(M)$ acts on this space, but the action is not diffeomorphism invariant. However, if M is 2-dimensional one may write $F = f \otimes \omega$ where ω is the volume form on M and f is a section of $P \times_{\text{Ad}} \mathbf{g}$, and then

$$S(A) = -\frac{1}{2} \int_M \text{Tr}(f^2) \, \omega.$$

It follows that the action $S(A)$ is invariant under the subgroup of diffeomorphisms preserving the volume form ω. So upon quantization one expects to—and does—obtain something analogous to a topological quantum field theory, but in which diffeomorphism invariance is replaced by invariance under this subgroup. Strictly speaking, then, many of the results of the previous section do not apply. In particular, this theory has an honest Hamiltonian, rather than a Hamiltonian constraint. Still, it illustrates some interesting aspects of gauge field/string field duality.

The Riemannian case of 2d Yang–Mills theory has been extensively investigated. An equation for the vacuum expectation values of Wilson loops for the theory on Euclidean $\mathbf{R}^2$ was found by Migdal [36], and these expectation values were explicitly calculated by Kazakov [32]. These calculations were made rigorous using stochastic differential equation techniques by L. Gross, King and Sengupta [28], as well as Driver [18]. The classical Yang–Mills equations on Riemann surfaces were extensively investigated by Atiyah and Bott [8], and the quantum theory on Riemann surfaces has been studied by Rusakov [51], Fine [19], and Witten [55]. In particular, Witten has shown that the quantization of 2d Yang–Mills theory gives a mathematical structure very close to that of a topological quantum field theory, with a Hilbert space $Z(S^1 \cup \cdots \cup S^1)$ associated to each compact 1-manifold $S^1 \cup \cdots \cup S^1$, and a vector $Z(M, \alpha) \in Z(\partial M)$ for each compact oriented 2-manifold M with boundary having total area $\alpha = \int_M \omega$. Similar results were also obtained by Blau and Thompson [11].

Here, however, it will be a bit simpler to discuss Yang–Mills theory on $\mathbf{R} \times S^1$ with the Lorentzian metric

$$dt^2 - dx^2,$$

where $t \in \mathbf{R}$, $x \in S^1$, as done by Rajeev [47]. This will simultaneously serve as a brief introduction to the idea of quantizing gauge theories after symplectic reduction, which will also be important in 3d quantum gravity. This approach is an alternative to the path integral approach of the previous section, and in some cases is easier to make rigorous.

Any G-bundle $P \to \mathbf{R} \times S^1$ is trivial, so we fix a trivialization and identify a connection on P with a $\mathbf{g}$-valued 1-form on $\mathbf{R} \times S^1$. The classical configuration space of the theory is the space $\mathcal{A}$ of connections on $P|_{\{0\} \times S^1}$. This may be identified with the space of $\mathbf{g}$-valued 1-forms on S^1. The classical phase space of the theory is the cotangent bundle $T^*\mathcal{A}$. Note that a tangent vector $v \in T_A\mathcal{A}$ may be identified with a $\mathbf{g}$-valued 1-form on S^1. We may also think of a $\mathbf{g}$-valued 1-form E on S^1 as a cotangent vector, using the non-degenerate inner product:

$$\langle E, v \rangle = - \int_{S^1} \text{Tr}(E \wedge \star v),$$

We thus regard the phase space $T^*\mathcal{A}$ as the space of pairs (A, E) of $\mathbf{g}$-valued 1-forms on S^1.

Given a connection on P solving the Yang–Mills equation we obtain a point (A, E) of the phase space $T^*\mathcal{A}$ as follows: let A be the pullback of the connection to $\{0\} \times S^1$, and let E be its covariant time derivative pulled back to $\{0\} \times S^1$. The pair (A, E) is called the initial data for the solution, and in physics A is called the vector potential and E the electric field. The Yang–Mills equation implies a constraint on (A, E), the Gauss law

$$d_A \star E = 0,$$

and any pair (A, E) satisfying this constraint are the initial data for some solution of the Yang–Mills equation. However, this solution is not unique, owing to the gauge invariance of the equation. Moreover, the loop group $\mathcal{G} = C^\infty(S^1, G)$ acts as gauge transformations on $\mathcal{A}$, and this action lifts naturally to an action on $T^*\mathcal{A}$, given by:

$$g\colon (A, E) \to (gAg^{-1} + g d(g^{-1}), gEg^{-1}).$$

Two points in the phase space $T^*\mathcal{A}$ are to be regarded as physically equivalent if they differ by a gauge transformation.

In this sort of situation it is natural to try to construct a smaller, more physically relevant 'reduced phase space' using the process of symplectic reduction. The phase space $T^*\mathcal{A}$ is a symplectic manifold, but the constraint subspace

$$\{(A, E)\mid d_A \star E = 0\} \subset T^*\mathcal{A}$$

is not. However, the constraint $d_A \star E$, integrated against any $f \in C^\infty(S^1, \mathbf{g})$ as follows,

$$\int_{S^1} \mathrm{Tr}(f d_A \star E),$$

gives a function on phase space that generates a Hamiltonian flow coinciding with a 1-parameter group of gauge transformations. In fact, all 1-parameter subgroups of $\mathcal{G}$ are generated by the constraint in this fashion. Consequently, the quotient of the constraint subspace by $\mathcal{G}$ is again a symplectic manifold, the reduced phase space.

In the case at hand there is a very concrete description of the reduced phase space. First, by basic results on moduli spaces of flat connections, the 'reduced configuration space' $\mathcal{A}/\mathcal{G}$ may be naturally identified with $\mathrm{Hom}(\pi_1(S^1), G)/\mathrm{Ad}G$, which is just $G/\mathrm{Ad}G$. Alternatively, one can see this quite concretely. We may first take the quotient of $\mathcal{A}$ by only those gauge transformations that equal the identity at a given point of S^1:

$$\mathcal{G}_0 = \{g \in C^\infty(S^1, G)\mid g(0) = 1\}.$$

This 'almost reduced' configuration space $\mathcal{A}/\mathcal{G}_0$ is diffeomorphic to G itself, with an explicit diffeomorphism taking each equivalence class $[A]$ to its holonomy around the circle:

$$[A] \mapsto \mathrm{P}\, e^{\oint_{S^1} A}.$$

The remaining gauge transformations form the group $\mathcal{G}/\mathcal{G}_0 \cong G$, which acts on the almost reduced configuration space G by conjugation, so $\mathcal{A}/\mathcal{G} \cong G/\mathrm{Ad}G$.

Next, writing $E = e\,dx$, the Gauss law says that $e \in C^\infty(S^1, \mathbf{g})$ is a flat section, and hence determined by its value at the basepoint of S^1. It follows that any point (A, E) in the constraint subspace is determined by $A \in \mathcal{A}$ together with $e(0) \in \mathbf{g}$. The quotient of the constraint subspace by $\mathcal{G}_0$, the 'almost reduced' phase space, is thus identified with T^*G. It follows that the quotient of the constraint subspace by all of $\mathcal{G}$, the reduced phase space, is identified with $T^*(G/\mathrm{Ad}G)$.

The advantage of the almost reduced configuration space and phase space is that they are manifolds. Observables of the classical theory can be identified either with functions on the reduced phase space, or functions on the almost reduced phase space T^*G that are constant on the orbits of the lift of the adjoint action of G. For example, the Yang-Mills Hamiltonian is initially a function on $T^*\mathcal{A}$:

$$H(A, E) = \frac{1}{2}\langle E, E\rangle$$

but by the process of symplectic reduction one obtains a corresponding Hamiltonian on the reduced phase space. One can, however, carry out only part of the process of symplectic reduction, and obtain a Hamiltonian function on the almost reduced phase space. This is just the Hamiltonian for a free particle on G, i.e. for any $p \in T^*_g G$ it is given by

$$H(g, p) = \frac{1}{2}\|p\|^2$$

with the obvious inner product on $T^*_g G$.

Now let us consider quantizing 2-dimensional Yang–Mills theory. What should be the Hilbert space for the quantized theory on $\mathbf{R} \times S^1$? As described in the previous section, it is natural to take L^2 of the reduced configuration space $\mathcal{A}/\mathcal{G}$. (Since the theory is not generally covariant, the diffeomorphism and Hamiltonian constraints do not enter; the 'kinematical' Hilbert space is the physical Hilbert space.) However, to define $L^2(\mathcal{A}/\mathcal{G})$ requires choosing a measure on $\mathcal{A}/\mathcal{G} = G/\mathrm{Ad}G$. We will choose the pushforward of normalized Haar measure on G by the quotient map $G \to G/\mathrm{Ad}G$. This measure has the advantage of mathematical elegance. While one could also argue for it on physical grounds, we prefer simply to show *ex post facto* that it gives an interesting quantum theory consistent with other approaches to 2d Yang–Mills theory.

To begin with, note that this measure gives a Hilbert space isomorphism

$$L^2(\mathcal{A}/\mathcal{G}) \cong L^2(G)_{inv}$$

where the right side denotes the subspace of $L^2(G)$ consisting of functions constant on each conjugacy class of G. Let χ_ρ denote the character of an equivalence class ρ of irreducible representations of G. Then by the Schur orthogonality relations, the set $\{\chi_\rho\}$ forms an orthonormal basis of $L^2(G)_{inv}$. In fact, the Hamiltonian of the quantum theory is diagonalized by this basis. Since the Yang–Mills Hamiltonian on the almost reduced phase space T^*G is that of a classical free particle on G, we take the quantum Hamiltonian to be that for a quantum free particle on G:

$$H = \Delta/2$$

where Δ is the (non-negative) Laplacian on G. When we decompose the regular representation of G into irreducibles, the function χ_ρ lies in the sum of copies of the representation ρ, so

$$H\chi_\rho = \frac{1}{2}c_2(\rho)\chi_\rho, \tag{3.1}$$

where $c_2(\rho)$ is the quadratic Casimir of G in the representation ρ. Note that the vacuum (the eigenvector of H with lowest eigenvalue) is the function 1, which is χ_ρ for ρ the trivial representation.

In a sense this diagonalization of the Hamiltonian completes the solution of Yang–Mills theory on $\mathbf{R} \times S^1$. However, extracting the physics from this solution requires computing expectation values of physically interesting observables. To take a step in this direction, and to make the connection to string theory, let us consider the Wilson loop observables. Recall that given a based loop $\gamma: S^1 \to S^1$, the classical Wilson loop $T(\gamma, A)$ is defined by

$$T(\gamma, A) = \mathrm{Tr}\mathrm{P}e^{\oint_\gamma A}.$$

We may think of $T(\gamma) = T(\gamma, \cdot)$ as a function on the reduced configuration space $\mathcal{A}/\mathcal{G}$, but it lifts to a function on the almost reduced configuration space G, and we prefer to think of it as such. In the case at hand these Wilson loop observables depend only on the homotopy class of the loop, because all connections on S^1 are flat. In the string field picture of Section 2, we obtain a theory in which all physical states have

$$\psi(\eta_1, \ldots, \eta_n) = \psi(\gamma_1, \ldots, \gamma_n)$$

when η_i is homotopic to γ_i for all i. We will see this again in 3d quantum gravity. Letting $\gamma_n: S^1 \to S^1$ be an arbitrary loop of winding number n, we have

$$T(\gamma_n, g) = \mathrm{Tr}(g^n).$$

Since the classical Wilson loop observables are functions on configuration space, we may quantize them by interpreting them as multiplication operators acting on $L^2(G)_{inv}$:

$$(T(\gamma_n)\psi)(g) = \mathrm{Tr}(g^n)\psi(g).$$

We can also form elements of $L^2(G)_{inv}$ by applying products of these operators to the vacuum. Let

$$|n_1, \ldots, n_k\rangle = T(\gamma_{n_1}) \cdots T(\gamma_{n_k})1.$$

The states $|n_1, \ldots, n_k\rangle$ may also be regarded as states of a string theory in which k strings are present, with winding numbers $n_1, \ldots, n_k$, respectively. For convenience, we define $|\emptyset\rangle$ to be the vacuum state.

The idea of describing the Yang–Mills Hamiltonian in terms of these string states goes back at least to Jevicki's work [30] on lattice gauge theory. More recently, string states appear prominently in the work of Gross, Taylor, Douglas, Minahan, Polychronakos, and others [17, 27, 37, 38, 39] on 2d Yang–Mills theory as a string theory. All these authors, however, work primarily with the large-N limit of the theory, for since the work of 't Hooft [52] it has been clear that $SU(N)$ Yang–Mills theory simplifies as $N \to \infty$. In what follows we will use many ideas from these authors, but give a string-theoretic formula for the $SU(N)$ Yang–Mills Hamiltonian that is exact for arbitrary N, instead of working in the large-N limit.

The resemblance of the 'string states' $|n_1, \ldots, n_k\rangle$ to states in a bosonic Fock space should be clear. In particular, the $T(\gamma_n)$ are analogous to 'creation operators'. However, we do *not* generally have a representation of the canonical commutation relations. In fact, the string states do not necessarily span $L^2(G)_{inv}$, although they do in some interesting cases. They are never linearly independent, because the Wilson loops satisfy relations. One always has $T(\gamma_0) = \text{Tr}(1)$, for example, and for any particular group G the Wilson loops will satisfy identities called Mandelstam identities. For example, for $G = SU(2)$ and taking traces in the fundamental representation, the Mandelstam identity is

$$T(\gamma_n)T(\gamma_m) = T(\gamma_{n+m}) + T(\gamma_{n-m}).$$

Note that this implies that

$$|n, m\rangle = |n + m\rangle + |n - m\rangle,$$

so the total number of strings present in a given state is ambiguous. In other words, there is no analog of the Fock space 'number operator' on $L^2(G)_{inv}$.

For the rest of this section we set $G = SU(N)$ and take traces in the fundamental representation. In this case the string states do span $L^2(G)_{inv}$, and all the linear dependencies between string states are consequences of the following Mandelstam identities [22]. Given loops $\eta_1, \ldots, \eta_k$ in S^1, let

$$M_k(\eta_1, \ldots, \eta_k) = \frac{1}{k!} \sum_{\sigma \in S_k} \text{sgn}(\sigma) T(g_{j_{11}} \cdots g_{j_{1n_1}}) \cdots T(g_{j_{k1}} \cdots g_{j_{kn_k}})$$

where σ has the cycle structure $(j_{11} \cdots j_{1n_1}) \cdots (j_{k1} \cdots j_{kn_k})$. Then

$$M_N(\eta, \ldots, \eta) = 1$$

for all loops η, and

$$M_{N+1}(\eta_1, \ldots, \eta_{N+1}) = 0$$

for all loops η_i. There are also explicit formulae expressing the string states in terms of the basis $\{\chi_\rho\}$ of characters. These formulae are based on the classical theory of Young diagrams, which we shall briefly review. The importance of this theory for 2-dimensional physics was stressed by

Nomura [42], and plays a primary role in Gross and Taylor's work on 2d Yang–Mills theory [27]. As we shall see, Young diagrams describe a 'duality' between the representation theory of $SU(N)$ and of the symmetric groups S_n which can be viewed as a mathematical reflection of string field/gauge field duality.

First, note using the Mandelstam identities that the string states

$$|n_1, \ldots, n_k\rangle$$

with all the n_i positive (but k possibly equal to zero) span $L^2(SU(N))_{inv}$. Thus we will restrict our attention for now to states of this kind, which we call 'right-handed'. There is a 1–1 correspondence between right-handed string states and conjugacy classes of permutations in symmetric groups, in which the string state $|n_1, \ldots, n_k\rangle$ corresponds to the conjugacy class σ of all permutations with cycles of length $n_1, \ldots, n_k$. Note that σ consists of permutations in $S_{n(\sigma)}$, where $n(\sigma) = n_1 + \cdots + n_k$. To take advantage of this correspondence, we simply define

$$|\sigma\rangle = |n_1, \ldots, n_k\rangle.$$

when σ is the conjugacy class of permutations with cycle lengths $n_1, \ldots, n_k$. We will assume without loss of generality that $n_1 \geq \cdots \geq n_k > 0$.

The rationale for this description of string states as conjugacy classes of permutations is in fact quite simple. Suppose we have length-minimizing strings in S^1 with winding numbers $n_1, \ldots, n_k$. Labelling each strand of string each time it crosses the point $x = 0$, for a total of $n = n_1 + \cdots + n_k$ labels, and following the strands around counterclockwise to $x = 2\pi$, we obtain a permutation of the labels, and hence an element of S_n. However, since the labelling was arbitrary, the string state really only defines a conjugacy class σ of elements of S_n.

In a Young diagram one draws a conjugacy class σ with cycles of length $n_1 \geq \cdots \geq n_k > 0$ as a diagram with k rows of boxes, having n_i boxes in the ith row. (See Fig. 2.) Let Y denote the set of Young diagrams.

On the one hand, there is a map from Young diagrams to equivalence classes of irreducible representations of $SU(N)$. Given $\rho \in Y$, we form an irreducible representation of $SU(N)$, which we also call ρ, by taking a tensor product of n copies of the fundamental representation, one copy for each box, and then antisymmetrizing over all copies in each column and symmetrizing over all copies in each row. This gives a 1–1 correspondence between Young diagrams with $< N$ rows and irreducible representations of $SU(N)$. If ρ has N rows it is equivalent to a representation coming from a Young diagram having $< N$ rows, and if ρ has $> N$ rows it is zero dimensional. We will write χ_ρ for the character of the representation ρ; if ρ has $> N$ rows $\chi_\rho = 0$.

On the other hand, Young diagrams with n boxes are in 1–1 correspon-

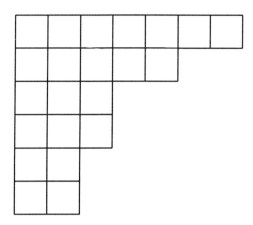

FIG. 2. Young diagram

dence with irreducible representations of S_n. This allows us to write the Frobenius relations expressing the string states $|\sigma\rangle$ in terms of characters χ_ρ and vice versa. Given $\rho \in Y$, we write $\tilde{\rho}$ for the corresponding representation of S_n. We define the function $\chi_{\tilde{\rho}}$ on S_n to be zero for $n(\rho) \neq n$, where $n(\rho)$ is the number of boxes in ρ. Then the Frobenius relations are

$$|\sigma\rangle = \sum_{\rho \in Y} \chi_{\tilde{\rho}}(\sigma)\chi_\rho, \tag{3.2}$$

and conversely

$$\chi_\rho = \frac{1}{n(\rho)!} \sum_{\sigma \in S_{n(\rho)}} \chi_{\tilde{\rho}}(\sigma) |\sigma\rangle. \tag{3.3}$$

The Yang–Mills Hamiltonian has a fairly simple description in terms of the basis of characters $\{\chi_\rho\}$. First, recall that eqn (3.1) expresses the Hamiltonian in terms of the Casimir. There is an explicit formula for the value of the $SU(N)$ Casimir in the representation ρ:

$$c_2(\rho) = Nn(\rho) - N^{-1}n(\rho)^2 + \frac{n(\rho)(n(\rho) - 1)\chi_{\tilde{\rho}}(\text{`2'})}{\dim(\tilde{\rho})}$$

where '2' denotes the conjugacy class of permutations in $S_{n(\rho)}$ with one cycle of length 2 and the rest of length 1. It follows that

$$H = \frac{1}{2}(NH_0 - N^{-1}H_0^2 + H_1) \tag{3.4}$$

where

$$H_0 \chi_\rho = n(\rho)\chi_\rho \qquad (3.5)$$

and

$$H_1 \chi_\rho = \frac{n(\rho)(n(\rho)-1)\chi_\rho(\text{`2'})}{\dim(\tilde\rho)}\chi_\rho. \qquad (3.6)$$

To express the operators H_0 and H_1 in string-theoretic terms, it is convenient to define string annihilation and creation operators satisfying the canonical commutation relations. As noted above, there is no natural way to do this in $L^2(SU(N))_{inv}$ since the string states are not linearly independent. The advantage of taking the '$N \to \infty$ limit' is that any finite set of distinct string states becomes linearly independent, in fact orthogonal, for sufficiently large N. We will proceed slightly differently, simply *defining* a space in which all the string states are independent. Let **H** be a Hilbert space having an orthonormal basis $\{\mathcal{X}_\rho\}_{\rho\in Y}$ indexed by all Young diagrams. For each $\sigma \in Y$, *define* a vector $|\sigma\rangle$ in **H** by the Frobenius relation (3.2). Then a calculation using the Schur orthogonality equations twice shows that these string states $|\sigma\rangle$ are not only linearly independent, but orthogonal:

$$\begin{aligned}
\langle\sigma|\sigma'\rangle &= \sum_{\rho,\rho'\in Y} \overline{\chi}_{\tilde\rho}(\sigma)\chi_{\tilde\rho'}(\sigma')\langle\mathcal{X}_\rho,\mathcal{X}_{\rho'}\rangle \\
&= \sum_{\rho\in Y} \overline{\chi}_{\tilde\rho}(\sigma)\chi_{\tilde\rho}(\sigma') \\
&= \frac{n(\sigma)!}{|\sigma|}\delta_{\sigma\sigma'}
\end{aligned}$$

where $|\sigma|$ is the number of elements in σ regarded as a conjugacy class in S_n. One can also *derive* the Frobenius relation (3.3) from these definitions and express the basis $\{\mathcal{X}_\rho\}$ in terms of the string states:

$$\mathcal{X}_\rho = \frac{1}{n(\rho)!} \sum_{\sigma\in S_{n(\rho)}} \chi_{\tilde\rho}(\sigma)|\sigma\rangle.$$

It follows that the string states form a basis for **H**.

The Yang–Mills Hilbert space $L^2(SU(N))_{inv}$ is essentially a quotient space of the string field Hilbert space **H**, with the (only densely defined) quotient map

$$j: \mathbf{H} \to L^2(SU(N))_{inv}$$

being given by

$$\mathcal{X}_\rho \mapsto \chi_\rho.$$

This quotient map sends the string state $|\sigma\rangle$ in **H** to the corresponding string state $|\sigma\rangle \in L^2(SU(N))_{inv}$. It follows that this quotient map is precisely that which identifies any two string states that are related by the Mandelstam identities. It was noted some time ago by Gliozzi and Virasoro [24] that Mandelstam identities on string states are strong evidence for a gauge field interpretation of a string field theory. Here in fact we will show that the Hamiltonian on the Yang–Mills Hilbert space $L^2(SU(N))_{inv}$ lifts to a Hamiltonian on **H** with a simple interpretation in terms of string interactions, so that 2-dimensional $SU(N)$ Yang–Mills theory is isomorphic to a quotient of a string theory by the Mandelstam identities. In the framework of the previous section, the Mandelstam identities would appear as part of the 'dynamical constraint' **K** of the string theory.

Following eqns (3.4–3.6), we define a Hamiltonian H on the string field Hilbert space **H** by

$$H = \frac{1}{2}(NH_0 - N^{-1}H_0^2 + H_1)$$

where

$$H_0 \mathcal{X}_\rho = n(\rho)\mathcal{X}_\rho, \qquad H_1 \mathcal{X}_\rho = \frac{n(n-1)\chi_{\bar{\rho}}(`2')}{\dim(\bar{\rho})}\mathcal{X}_\rho.$$

This clearly has the property that

$$Hj = jH,$$

so the Yang–Mills dynamics is the quotient of the string field dynamics. On **H** we can introduce creation operators a_j^* $(j > 0)$ by

$$a_j^*|n_1, \ldots, n_k\rangle = |j, n_1, \ldots, n_k\rangle,$$

and define the annihilation operator a_j to be the adjoint of a_j^*. These satisfy the following commutation relations:

$$[a_j, a_k] = [a_j^*, a_k^*] = 0, \qquad [a_j, a_k^*] = j\delta_{jk}.$$

We could eliminate the factor of j and obtain the usual canonical commutation relations by a simple rescaling, but it is more convenient not to. We then claim that

$$H_0 = \sum_{j>0} a_j^* a_j$$

and

$$H_1 = \sum_{j,k>0} a_{j+k}^* a_j a_k + a_j^* a_k^* a_{j+k}.$$

These follow from calculations which we briefly sketch here. The Frobenius relations and the definition of H_0 give

$$H_0|\sigma\rangle = n(\sigma)|\sigma\rangle, \tag{3.7}$$

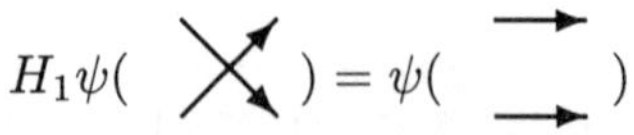

$$H_1\psi(\ \ ) = \psi(\ \ )$$

Fig. 3. Two-string interaction in 1-dimensional space

and this implies the formula for H_0 as a sum of harmonic oscillator Hamiltonians $a_j^* a_j$. Similarly, the Frobenius relations and the definition of H_1 give

$$H_1|\sigma\rangle = \sum_{\rho \in Y} \frac{n(\sigma)(n(\sigma) - 1)}{\dim(\tilde{\rho})} \chi_{\tilde{\rho}}(`2')\chi_{\tilde{\rho}}(\sigma)\chi_\rho.$$

Since there are $n(n-1)/2$ permutations $\tau \in S_{n(\sigma)}$ lying in the conjugacy class '2', we may rewrite this as

$$H_1|\sigma\rangle = \sum_{\rho \in Y, \tau \in `2'} \frac{2}{\dim(\tilde{\rho})} \chi_{\tilde{\rho}}(\sigma)\chi_{\tilde{\rho}}(\tau)\chi_\rho.$$

Since

$$\sum_{\tau \in `2'} \frac{1}{\dim(\tilde{\rho})} \chi_{\tilde{\rho}}(\sigma)\chi_{\tilde{\rho}}(\tau) = \sum_{\tau \in `2'} \chi_{\tilde{\rho}}(\sigma\tau)$$

the Frobenius relations give

$$H_1|\sigma\rangle = 2 \sum_{\tau \in `2'} |\sigma\tau\rangle. \tag{3.8}$$

An analysis of the effect of composing σ with all possible $\tau \in `2'$ shows that either one cycle of σ will be broken into two cycles, or two will be joined to form one, giving the expression above for H_1 in terms of annihilation and creation operators.

We may interpret the Hamiltonian in terms of strings as follows. By eqn (3.7), H_0 can be regarded as a 'string tension' term, since if we represent a string state $|n_1, \ldots, n_k\rangle$ by length-minimizing loops, it is an eigenvector of H_0 with eigenvalue equal to $n_1 + \cdots + n_k$, proportional to the sum of the lengths of the loops.

By eqn (3.8), H_1 corresponds to a two-string interaction as in Fig. 3. In this figure only the x coordinate is to be taken seriously; the other has been introduced only to keep track of the identities of the strings. Also, we have switched to treating states as functions on the space of multiloops. As the figure indicates, this kind of interaction is a 1-dimensional version of that which gave the HOMFLY invariant of links in 3-dimensional space in the previous section. Here, however, we have a true Hamiltonian rather than a Hamiltonian constraint.

Figure 3 can also be regarded as two frames of a 'movie' of a string worldsheet in 2-dimensional spacetime. Similar movies have been used by Carter and Saito to describe string worldsheets in 4-dimensional spacetime

[14]. If we draw the string worldsheet corresponding to this movie we obtain a surface with a branch point. Indeed, in the path integral approach of Gross and Taylor this kind of term appears in the partition function as part of a sum over string histories, associated to those histories with branch points. They also show that the H_0^2 term corresponds to string worldsheets with handles. When considering the $1/N$ expansion of the theory, it is convenient to divide the Hamiltonian H by N, so that it converges to H_0 as $N \to \infty$. Then the H_0^2 term is proportional to $1/N^2$. This is in accord with the observation by 't Hooft [52] that in an expansion of the free energy (logarithm of the partition function) as a power series in $1/N$, string worldsheets of genus g give terms proportional to $1/N^{2-2g}$.

From the work of Gross and Taylor it is also clear that in addition to the space **H** spanned by right-handed string states one should also consider a space with a basis of 'left-handed' string states $|n_1, \ldots, n_k\rangle$ with $n_i < 0$. The total Hilbert space of the string theory is then the tensor product $\mathbf{H_+} \otimes \mathbf{H_-}$ of right-handed and left-handed state spaces. This does not describe any new states in the Yang–Mills theory *per se*, but it is more natural from the string-theoretic point of view. It follows from the work of Minahan and Polychronakos that there is a Hamiltonian H on $\mathbf{H_+} \otimes \mathbf{H_-}$ naturally described in terms of string interactions and a densely defined quotient map $j : \mathbf{H_+} \otimes \mathbf{H_-} \to L^2(SU(N))_{inv}$ such that $Hj = jH$.

4 Quantum gravity in 3 dimensions

Now let us turn to a more sophisticated model, 3-dimensional quantum gravity. In 3 dimensions, Einstein's equations say simply that the spacetime metric is flat, so there are no local degrees of freedom. The theory is therefore only interesting on topologically non-trivial spacetimes. Interest in the mathematics of this theory increased when Witten [2] reformulated it as a Chern–Simons theory. Since then, many approaches to the subject have been developed, not all equivalent [12]. We will follow Ashtekar, Husain, Rovelli, Samuel and Smolin [1, 6] and treat 3-dimensional quantum gravity using the 'new variables' and the loop transform, and indicate some possible relations to string theory. It is important to note that there are some technical problems with the loop transform in Lorentzian quantum gravity, since the gauge group is then non-compact [35]. These are presently being addressed by Ashtekar and Loll [5] in the 3-dimensional case, but for simplicity of presentation we will sidestep them by working with the Riemannian case, where the gauge group is $SO(3)$.

It is easiest to describe the various action principles for gravity using the abstract index notation popular in general relativity, but we will instead translate them into language that may be more familiar to mathematicians, since this seems not to have been done yet. In this section we describe the 'Witten action', applicable to the 3-dimensional case; in the next section we describe the 'Palatini action', which applies to any dimension, and the 'Ashtekar action', which applies to 4 dimensions. The relationship between these action principles has been discussed rather thoroughly by Peldan [43].

Let the spacetime M be an orientable 3-manifold. Fix a real vector bundle T over M that is isomorphic to—but *not* canonically identified with—the tangent bundle TM, and fix a Riemannian metric η and an orientation on T. These define a 'volume form' ϵ on T, that is a nowhere-vanishing section of $\Lambda^3 T^*$. The basic fields of the theory are then taken to be a metric-preserving connection A on T, or '$SO(3)$ connection', together with a T-valued 1-form e on M. Using the isomorphism $T \cong T^*$ given by the metric, the curvature F of A may be identified with a $\Lambda^2 T$-valued 2-form. It follows that the wedge product $e \wedge F$ may be defined as a $\Lambda^3 T$-valued 3-form. Pairing this with ϵ to obtain an ordinary 3-form and then integrating over spacetime, we obtain the Witten action

$$S(A, e) = \frac{1}{2} \int_M \epsilon(e \wedge F).$$

The classical equations of motion obtained by extremizing this action are

$$F = 0$$

and

$$d_A e = 0.$$

Note that we can pull back the metric η on E by $e \colon TM \to T$ to obtain a 'Riemannian metric' on M, which, however, is only non-degenerate when e is an isomorphism. When e is an isomorphism we can also use it to pull back the connection to a metric-preserving connection on TM. In this case, the equations of motion say simply that this connection is the Levi-Civita connection of the metric on M, and that the metric on M is flat. The formalism involving the fields A and e can thus be regarded as a device for extending the usual Einstein equations in 3 dimensions to the case of degenerate 'metrics' on M.

Now suppose that $M = \mathbf{R} \times X$, where X is a compact oriented 2-manifold. The classical configuration and phase spaces and their reduction by gauge transformations are reminiscent of those for 2d Yang-Mills theory. There are, however, a number of subtleties, and we only present the final results. The classical configuration space can be taken as the space $\mathcal{A}$ of metric-preserving connections on $T|X$, which we call $SO(3)$ connections on X. The classical phase space is then the cotangent bundle $T^*\mathcal{A}$. Note that a tangent vector $v \in T_A\mathcal{A}$ is a $\Lambda^2 T$-valued 1-form on X. We can thus regard a T-valued 1-form $\tilde{E}$ on X as a cotangent vector by means of the pairing

$$\tilde{E}(v) = \int_X \epsilon(\tilde{E} \wedge v).$$

Thus given any solution (A, e) of the classical equations of motion, we can pull back A and e to the surface $\{0\} \times X$ and get an $SO(3)$ connection and a $\mathcal{T}$-valued 1-form on X, that is a point in the phase space $T^*\mathcal{A}$. This is usually written $(A, \tilde{E})$, where $\tilde{E}$ plays a role analogous to the electric field in Yang–Mills theory.

The classical equations of motion imply constraints on $(A, \tilde{E}) \in T^*\mathcal{A}$ which define a reduced phase space. These are the Gauss law, which in this context is

$$d_A \tilde{E} = 0,$$

and the vanishing of the curvature B of the connection A on $T|X$, which is analogous to the magnetic field:

$$B = 0.$$

The latter constraint subsumes both the diffeomorphism and Hamiltonian constraints of the theory. The reduced phase space for the theory turns out to be $T^*(\mathcal{A}_0/\mathcal{G})$, where $\mathcal{A}_0$ is the space of flat $SO(3)$ connections on X, and $\mathcal{G}$ is the group of gauge transformations [6]. As in 2d Yang–Mills theory, it will be attractive to quantize after imposing constraints, taking the physical state space of the quantized theory to be L^2 of the reduced configuration space, if we can find a tractable description of $\mathcal{A}_0/\mathcal{G}$.

A quite concrete description of $\mathcal{A}_0/\mathcal{G}$ was given by Goldman [25]. The moduli space $\mathcal{F}$ of flat $SO(3)$-bundles has two connected components, corresponding to the two isomorphism classes of $SO(3)$ bundles on M. The component corresponding to the bundle $T|X$ is precisely the space $\mathcal{A}_0/\mathcal{G}$, so we wish to describe this component.

There is a natural identification

$$\mathcal{F} \cong \mathrm{Hom}(\pi_1(X), SO(3))/\mathrm{Ad}(SO(3)),$$

given by associating to any flat bundle the holonomies around (homotopy classes of) loops. Suppose that X has genus g. Then the group $\pi_1(X)$ has a presentation with $2g$ generators $x_1, y_1, \ldots, x_g, y_g$ satisfying the relation

$$R(x_i, y_i) = (x_1 y_1 x_1^{-1} y_1^{-1}) \cdots (x_g y_g x_g^{-1} y_g^{-1}) = 1.$$

An element of $\mathrm{Hom}(\pi_1(X), SO(3))$ may thus be identified with a collection $u_1, v_1, \ldots, u_g, v_g$ of elements of $SO(3)$, satisfying

$$R(u_i, v_i) = 1,$$

and a point in $\mathcal{F}$ is an equivalence class $[u_i, v_i]$ of such collections.

The two isomorphism classes of $SO(3)$ bundles on M are distinguished by their second Stiefel–Whitney number $w_2 \in \mathbf{Z}_2$. The bundle $T|X$ is trivial so $w_2(T|X) = 0$ We can calculate w_2 for any point $[u_i, v_i] \in \mathcal{F}$ by the following method. For all the elements $u_i, v_i \in SO(3)$, choose lifts $\tilde{u}_i, \tilde{v}_i$

to the universal cover $\widetilde{SO}(3) \cong SU(2)$. Then

$$(-1)^{w_2} = R(\tilde{u}_i, \tilde{v}_i).$$

It follows that we may think of points of $\mathcal{A}_0/\mathcal{G}$ as equivalence classes of $2g$-tuples (u_i, v_i) of elements of $SO(3)$ admitting lifts $\tilde{u}_i, \tilde{v}_i$ with

$$R(\tilde{u}_i, \tilde{v}_i) = 1,$$

where the equivalence relation is given by the adjoint action of $SO(3)$.

In fact $\mathcal{A}_0/\mathcal{G}$ is not a manifold, but a singular variety. This has been investigated by Narasimhan and Seshadri [41], and shown to be dimension $d = 6g - 6$ for $g \geq 2$, or $d = 2$ for $g = 1$ (the case $g = 0$ is trivial and will be excluded below). As noted, it is natural to take $L^2(\mathcal{A}_0/\mathcal{G})$ to be the physical state space, but to define this one must choose a measure on $\mathcal{A}_0/\mathcal{G}$. As noted by Goldman [25], there is a symplectic structure Ω on $\mathcal{A}_0/\mathcal{G}$ coming from the following 2-form on $\mathcal{A}_0$:

$$\Omega(B, C) = \int_X \mathrm{Tr}(B \wedge C),$$

in which we identify the tangent vectors B, C with $\mathrm{End}(T|X)$-valued 1-forms. The d-fold wedge product $\Omega \wedge \cdots \wedge \Omega$ defines a measure μ on $\mathcal{A}_0/\mathcal{G}$, the Liouville measure. On the grounds of elegance and diffeomorphism invariance it is customary to use this measure to define the physical state space $L^2(\mathcal{A}_0/\mathcal{G})$.

It would be satisfying if there were a string-theoretic interpretation of the inner product in $L^2(\mathcal{A}_0/\mathcal{G})$ along the lines of Section 2. Note that we may define 'string states' in this space as follows. Given any loop γ in X, the Wilson loop observable $T(\gamma)$ is a multiplication operator on $L^2(\mathcal{A}_0/\mathcal{G})$ that only depends on the homotopy class of γ. As in the case of 2d Yang–Mills theory, we can form elements of $L^2(\mathcal{A}_0/\mathcal{G})$ by applying products of these operators to the function 1, so given $\gamma_1, \ldots, \gamma_k \in \pi_1(X)$, define

$$|\gamma_1, \ldots, \gamma_k\rangle = T(\gamma_1) \cdots T(\gamma_k)1.$$

The first step towards a string-theoretic interpretation of 3d quantum gravity would be a formula for an inner product of string states

$$\langle \gamma_1, \ldots, \gamma_k | \gamma_1', \ldots, \gamma_{k'}' \rangle,$$

which is just a particular kind of integral over $\mathcal{A}_0/\mathcal{G}$. Note that this sort of integral makes sense when $\mathcal{A}_0/\mathcal{G}$ is the moduli space of flat connections for a trivial $SO(N)$ bundle over X, for any N. Alternatively, one could formulate 3d quantum gravity as a theory of $SU(2)$ connections and then generalize to $SU(N)$.

In fact, it appears that this sort of integral can be computed using 2d Yang–Mills theory on a Riemann surface, by introducing a coupling constant λ in the action:

$$S(A) = -\frac{1}{2\lambda^2} \int_M \mathrm{Tr}(F \wedge \star F),$$

computing the appropriate vacuum expectation value of Wilson loops, and then taking the limit $\lambda \to 0$. This procedure has been discussed by Blau and Thompson [11] and Witten [55]. One thus expects that this sort of integral simplifies in the $N \to \infty$ limit just as it does in 2d Yang–Mills theory, giving a formula for

$$\langle \gamma_1, \ldots, \gamma_k | \gamma_1', \ldots, \gamma_{k'}' \rangle$$

as a sum over ambient isotopy classes of surfaces $f \colon \Sigma \to [0,t] \times X$ having the loops γ_i, γ_i' as boundaries. This, together with a method of treating the finite N case by imposing Mandelstam identities, would give a string-theoretic interpretation of 3d quantum gravity. Taylor and the author are currently trying to work out the details of this program.

Before concluding this section, it is worth noting another generally covariant gauge theory in 3 dimensions, Chern–Simons theory. Here one fixes an arbitrary Lie group G and a G-bundle $P \to M$ over spacetime, and the field of the theory is a connection A on P. The action is given by

$$S(A) = \frac{k}{4\pi} \int \mathrm{Tr}\left(A \wedge dA + \frac{2}{3} A \wedge A \wedge A\right).$$

As noted by Witten [2], 3d quantum gravity as we have described it is essentially the same Chern–Simons theory with gauge group $ISO(3)$, the Euclidean group in 3 dimension, with the $SO(3)$ connection and triad field appearing as two parts of an $ISO(3)$ connection. There is a profound connection between Chern–Simons theory and knot theory, first demonstrated by Witten [54], and then elaborated by many researchers (see, for example, [7]). This theory does not quite fit our formalism because in it the space $\mathcal{A}_0/\mathcal{G}$ of flat connections modulo gauge transformations plays the role of a phase space, with the Goldman symplectic structure, rather than a configuration space. Nonetheless, there are a number of clues that Chern–Simons theory admits a reformulation as a generally covariant string field theory. In fact, Witten has given such an interpretation using open strings and the Batalin–Vilkovisky formalism [56]. Moreover, for the gauge groups $SU(N)$ Periwal has expressed the partition function for Chern–Simons theory on S^3, in the $N \to \infty$ limit, in terms of integrals over moduli spaces of Riemann surfaces. In the case $N = 2$ there is also, as one would expect, an expression for the vacuum expectation value of Wilson loops, at least for the case of a link (where it is just the Kauffman bracket invariant), in terms of a sum over surfaces having that link as boundary [13]. It would be very worth while to reformulate Chern–Simons theory as a string theory at the level of elegance with which one can do so for 2d Yang–Mills theory, but this has not yet been done.

5   Quantum gravity in 4 dimensions

We begin by describing the Palatini and Ashtekar actions for general relativity. As in the previous section, we will sidestep certain problems with the loop transform by working with Riemannian rather than Lorentzian gravity. We shall then discuss some recent work on making the loop representation rigorous in this case, and indicate some mathematical issues that need to be explored to arrive at a string-theoretic interpretation of the theory.

Let the spacetime M be an orientable n-manifold. Fix a bundle $\mathcal{T}$ over M that is isomorphic to TM, and fix a Riemannian metric η and orientation on $\mathcal{T}$. These define a nowhere-vanishing section ϵ of $\Lambda^n \mathcal{T}^*$. The basic fields of the theory are then taken to be a metric-preserving connection A on $\mathcal{T}$, or '$SO(n)$ connection', and a $\mathcal{T}$-valued 1-form e. We require, however, that $e: TM \to \mathcal{T}$ be a bundle isomorphism; its inverse is called a 'frame field'. The metric η defines an isomorphism $\mathcal{T} \cong \mathcal{T}^*$ and allows us to identify the curvature F of A with a section of the bundle

$$\Lambda^2 \mathcal{T} \otimes \Lambda^2 T^* M.$$

We may also regard e^{-1} as a section of $\mathcal{T}^* \otimes TM$ and define $e^{-1} \wedge e^{-1}$ in the obvious manner as a section of the bundle

$$\Lambda^2 \mathcal{T}^* \otimes \Lambda^2 TM.$$

The natural pairing between these bundles gives rise to a function $F(e^{-1} \wedge e^{-1})$ on M. Using the isomorphism e, we can push forward ϵ to a volume form ω on M. The Palatini action for Riemannian gravity is then

$$S(A, e) = \frac{1}{2} \int_M F(e^{-1} \wedge e^{-1}) \, \omega.$$

We may use the isomorphism e to transfer the metric η and connection A to a metric and connection on the tangent bundle. Then the classical equations of motion derived from the Palatini action say precisely that this connection is the Levi-Civita connection of the metric, and that the metric satisfies the vacuum Einstein equations (i.e. is Ricci flat).

In 3 dimensions, the Palatini action reduces to the Witten action, which, however, is expressed in terms of e rather than e^{-1}. In 4 dimensions the Palatini action can be rewritten in a somewhat similar form. Namely, the wedge product $e \wedge e \wedge F$ is a $\Lambda^4 \mathcal{T}$-valued 4-form, and pairing it with ϵ to obtain an ordinary 4-form we have

$$S(A, e) = \frac{1}{2} \int_M \epsilon(e \wedge e \wedge F).$$

The Ashtekar action depends upon the fact that in 4 dimensions the metric and orientation on $\mathcal{T}$ define a Hodge star operator

$$*: \Lambda^2 \mathcal{T} \to \Lambda^2 \mathcal{T}$$

with $*^2 = 1$ (not to be confused with the Hodge star operator on differential forms). This allows us to write F as a sum $F_+ + F_-$ of self-dual and anti-self-dual parts:

$$*F_\pm = \pm F_\pm.$$

The remarkable fact is that the action

$$S(A, e) = \frac{1}{2} \int_M \epsilon(e \wedge e \wedge F_+)$$

gives the same equations of motion as the Palatini action. Moreover, suppose $\mathcal{T}$ is trivial, as is automatically the case when $M \cong \mathbf{R} \times X$. Then F is just an $\mathbf{so}(4)$-valued 2-form on M, and its decomposition into self-dual and anti-self-dual parts corresponds to the decomposition $\mathbf{so}(4) \cong \mathbf{so}(3) \oplus \mathbf{so}(3)$. Similarly, A is an $\mathbf{so}(4)$-valued 1-form, and may thus be written as a sum $A_+ + A_-$ of 'self-dual' and 'anti-self-dual' connections, which are 1-forms having values in the two copies of $\mathbf{so}(3)$. It is easy to see that F_+ is the curvature of A_+. This allows us to regard general relativity as the theory of a self-dual connection A_+ and a $\mathcal{T}$-valued 1-form e—the so-called 'new variables'—with the Ashtekar action

$$S(A_+, e) = \frac{1}{2} \int_M \epsilon(e \wedge e \wedge F_+).$$

Now suppose that $M = \mathbf{R} \times X$, where X is a compact oriented 3-manifold. We can take the classical configuration space to be space $\mathcal{A}$ of right-handed connections on $\mathcal{T}|X$, or equivalently (fixing a trivialization of $\mathcal{T}$), $\mathbf{so}(3)$-valued 1-forms on X. A tangent vector $v \in T_A\mathcal{A}$ is thus an $\mathbf{so}(3)$-valued 1-form, and an $\mathbf{so}(3)$-valued 2-form $\tilde{E}$ defines a cotangent vector by the pairing

$$\tilde{E}(v) = \int_X \mathrm{Tr}(\tilde{E} \wedge v).$$

A point in the classical phase space $T^*\mathcal{A}$ is thus a pair $(A, \tilde{E})$ consisting of an $\mathbf{so}(3)$-valued 1-form A and an $\mathbf{so}(3)$-valued 2-form $\tilde{E}$ on X. In the physics literature it is more common to use the natural isomorphism

$$\Lambda^2 T^* X \cong TX \otimes \Lambda^3 T^* X$$

given by the interior product to regard the 'gravitational electric field' $\tilde{E}$ as an $\mathbf{so}(3)$-valued vector density, that is, a section of $\mathbf{so}(3) \otimes TX \otimes \Lambda^3 T^* X$.

A solution (A_+, e) of the classical equations of motion determines a point $(A, \tilde{E}) \in T^*\mathcal{A}$ as follows. The 'gravitational vector potential' A is simply the pullback of A_+ to the surface $\{0\} \times X$. Obtaining $\tilde{E}$ from e is a somewhat subtler affair. First, split the bundle $\mathcal{T}$ as the direct sum of a 3-dimensional bundle $^3\mathcal{T}$ and a line bundle. By restricting to TX and then projecting down to $^3\mathcal{T}|X$, the map

$$e: TM \to \mathcal{T}$$

gives a map

$$^3e: TX \to {}^3\mathcal{T}|X,$$

called a 'cotriad field' on X. Since there is a natural isomorphism of the fibers of $^3\mathcal{T}$ with $\mathbf{so}(3)$, we may also regard this as an $\mathbf{so}(3)$-valued 1-form on X. Applying the Hodge star operator we obtain the $\mathbf{so}(3)$-valued 2-form $\tilde{E}$.

The classical equations of motion imply constraints on $(A, \tilde{E}) \in T^*\mathcal{A}$. These are the Gauss law

$$d_A \tilde{E} = 0,$$

and the diffeomorphism and Hamiltonian constraints. The latter two are most easily expressed if we treat $\tilde{E}$ as an $\mathbf{so}(3)$-valued vector density. Letting B denote the 'gravitational magnetic field', or curvature of the connection A, the diffeomorphism constraint is given by

$$\text{Tr}\, i_{\tilde{E}} B = 0$$

and the Hamiltonian constraint is given by

$$\text{Tr}\, i_{\tilde{E}} i_{\tilde{E}} B = 0.$$

Here the interior product $i_{\tilde{E}} B$ is defined using 3×3 matrix multiplication and is an $M_3(\mathbf{R}) \otimes \Lambda^3 T^* X$-valued 1-form; similarly, $i_{\tilde{E}} i_{\tilde{E}} B$ is an $M_3(\mathbf{R}) \otimes \Lambda^3 T^* X \otimes \Lambda^3 T^* X$-valued function.

In 2d Yang–Mills theory and 3d quantum gravity one can impose enough constraints before quantizing to obtain a finite-dimensional reduced configuration space, namely the space $\mathcal{A}_0/\mathcal{G}$ of flat connections modulo gauge transformations. In 4d quantum gravity this is no longer the case, so a more sophisticated strategy, first devised by Rovelli and Smolin [49], is required. Let us first sketch this without mentioning the formidable technical problems. The Gauss law constraint generates gauge transformations, so one forms the reduced phase space $T^*(\mathcal{A}/\mathcal{G})$. Quantizing, one obtains the kinematical Hilbert space $\mathbf{H}_{kin} = L^2(\mathcal{A}/\mathcal{G})$. One then applies the loop transform and takes $\mathbf{H}_{kin} = \text{Fun}(\mathcal{M})$ to be a space of functions of multiloops in X. The diffeomorphism constraint generates the action of $\text{Diff}_0(X)$ on $\mathcal{A}/\mathcal{G}$, so in the quantum theory one takes $\mathbf{H}_{diff}$ to be the subspace of $\text{Diff}_0(X)$-invariant elements of $\text{Fun}(\mathcal{M})$. One may then either attempt to represent the Hamiltonian constraint as operators on $\mathbf{H}_{kin}$, and define the image of their common kernel in $\mathbf{H}_{diff}$ to be the physical state space $\mathbf{H}_{phys}$, or attempt to represent the Hamiltonian constraint directly as operators on $\mathbf{H}_{diff}$ and define the kernel to be $\mathbf{H}_{phys}$. (The latter approach is still under development by Rovelli and Smolin [50].)

Even at this formal level, the full space $\mathbf{H}_{phys}$ has not yet been determined. In their original work, Rovelli and Smolin [49] obtained a large set of physical states corresponding to ambient isotopy classes of links in X. More recently, physical states have been constructed from familiar link invariants such as the Kauffman bracket and certain coefficients of the Alexander polynomial to all of $\mathcal{M}$. Some recent developments along these lines have been reviewed by Pullin [46]. This approach makes use of the connection between 4d quantum gravity with cosmological constant and Chern–Simons theory in 3 dimensions. It is this work that suggests a profound connection between knot theory and quantum gravity.

There are, however, significant problems with turning all of this work into rigorous mathematics, so at this point we shall return to where we left off in Section 2 and discuss some of the difficulties. In Section 2 we were quite naive concerning many details of analysis—deliberately so, to indicate the basic ideas without becoming immersed in technicalities. In particular, one does not really expect to have interesting diffeomorphism-invariant measures on the space $\mathcal{A}/\mathcal{G}$ of connections modulo gauge transformations in this case. At best, one expects the existence of "generalized measures" sufficient for integrating a limited class of functions.

In fact, it is possible to go a certain distance without becoming involved with these considerations. In particular, the loop transform can be rigorously defined without fixing a measure or generalized measure on $\mathcal{A}/\mathcal{G}$ if one uses, not the Hilbert space formalism of the previous section, but a C*-algebraic formalism. A C*-algebra is an algebra A over the complex numbers with a norm and an adjoint or $*$ operation satisfying

$$(a^*)^* = a, \quad (\lambda a)^* = \bar{\lambda} a^*, \quad (a+b)^* = a^* + b^*, \quad (ab)^* = b^* a^*,$$

$$\|ab\| \le \|a\|\|b\|, \quad \|a^*a\| = \|a\|^2$$

for all a, b in the algebra and $\lambda \in \mathbf{C}$. In the C*-algebraic approach to physics, observables are represented by self-adjoint elements of A, while states are elements μ of the dual A^* that are positive, $\mu(a^*a) \ge 0$, and normalized, $\mu(1) = 1$. The number $\mu(a)$ then represents the expectation value of the observable a in the state μ. The relation to the more traditional Hilbert space approach to quantum physics is given by the Gelfand–Naimark–Segal (GNS) construction. Namely, a state μ on A defines an 'inner product' that may, however, be degenerate:

$$\langle a, b \rangle = \mu(a^*b).$$

Let $I \subseteq A$ denote the subspace of norm-zero states. Then A/I has an honest inner product and we let $\mathbf{H}$ denote the Hilbert space completion of A/I in the corresponding norm. It is then easy to check that I is a left ideal of A, so that A acts by left multiplication on A/I, and that this action extends uniquely to a representation of A as bounded linear operators on **H**. In particular, observables in A give rise to self-adjoint operators on **H**.

A C*-algebraic approach to the loop transform and generalized measures on $\mathcal{A}/\mathcal{G}$ was introduced by Ashtekar and Isham [3] in the context of $SU(2)$ gauge theory, and subsequently developed by Ashtekar, Lewandowski, and the author [4, 10]. The basic concept is that of the holonomy C*-algebra. Let X be a manifold, 'space', and let $P \to X$ be a principal G-bundle over X. Let $\mathcal{A}$ denote the space of smooth connections on P, and $\mathcal{G}$ the group of smooth gauge transformations. Fix a finite-dimensional representation ρ of G and define Wilson loop functions $T(\gamma) = T(\gamma, \cdot)$ on $\mathcal{A}/\mathcal{G}$ taking traces in this representation.

Define the 'holonomy algebra' to be the algebra of functions on $\mathcal{A}/\mathcal{G}$ generated by the functions $T(\gamma) = T(\gamma, \cdot)$. If we assume that G is compact and ρ is unitary, the functions $T(\gamma)$ are bounded and continuous (in the C^∞ topology on $\mathcal{A}/\mathcal{G}$). Moreover, the pointwise complex conjugate $T(\gamma)^*$ equals $T(\gamma^{-1})$, where γ^{-1} is the orientation-reversed loop. We may thus complete the holonomy algebra in the sup norm topology:

$$\|f\|_\infty = \sup_{A \in \mathcal{A}/\mathcal{G}} |f(A)|$$

and obtain a C*-algebra of bounded continuous functions on $\mathcal{A}/\mathcal{G}$, the 'holonomy C*-algebra', which we denote as $\mathrm{Fun}(\mathcal{A}/\mathcal{G})$ in order to make clear the relation to the previous section.

While in what follows we will assume that G is compact and ρ is unitary, it is important to emphasize that for Lorentzian quantum gravity G is not compact! This presents important problems in the loop representation of both 3- and 4-dimensional quantum gravity. Some progress in solving these problems has recently been made by Ashtekar, Lewandowski, and Loll [4, 5, 33].

Recall that in the previous section the loop transform of functions on $\mathcal{A}/\mathcal{G}$ was defined using a measure on $\mathcal{A}/\mathcal{G}$. It turns out to be more natural to define the loop transform not on $\mathrm{Fun}(\mathcal{A}/\mathcal{G})$ but on its dual, as this involves no arbitrary choices. Given $\mu \in \mathrm{Fun}(\mathcal{A}/\mathcal{G})^*$ we define its loop transform $\hat{\mu}$ to be the function on the space $\mathcal{M}$ of multiloops given by

$$\hat{\mu}(\gamma_1, \ldots, \gamma_n) = \mu(T(\gamma_1) \cdots T(\gamma_n)).$$

Let $\mathrm{Fun}(\mathcal{M})$ denote the range of the loop transform. In favorable cases, such as $G = SU(N)$ and ρ the fundamental representation, the loop transform is one-to-one, so

$$\mathrm{Fun}(\mathcal{A}/\mathcal{G})^* \cong \mathrm{Fun}(\mathcal{M}).$$

This is the real justification for the term 'string field/gauge field duality'.

We may take the 'generalized measures' on $\mathcal{A}/\mathcal{G}$ to be simply elements $\mu \in \text{Fun}(\mathcal{A}/\mathcal{G})^*$, thinking of the pairing $\mu(f)$ as the integral of $f \in \text{Fun}(\mathcal{A}/\mathcal{G})$. If μ is a state on $\text{Fun}(\mathcal{A}/\mathcal{G})$, we may construct the kinematical Hilbert space $\mathbf{H}_{kin}$ using the GNS construction. Note that the kinematical inner product

$$\langle [f], [g] \rangle_{kin} = \mu(f^*g)$$

then generalizes the L^2 inner product used in the previous section. Note that a choice of generalized measure μ also allows us to define the loop transform as a linear map from $\text{Fun}(\mathcal{A}/\mathcal{G})$ to $\text{Fun}(\mathcal{M})$

$$\hat{f}(\gamma_1, \ldots, \gamma_n) = \mu(T(\gamma_1) \cdots T(\gamma_n)f)$$

in a manner generalizing that of the previous section. Moreover, there is a unique inner product on $\text{Fun}(\mathcal{M})$ such that this map extends to a map from $\mathbf{H}_{kin}$ to the Hilbert space completion of $\text{Fun}(\mathcal{M})$. Note also that $\text{Diff}_0(X)$ acts on $\text{Fun}(\mathcal{A}/\mathcal{G})$ and dually on $\text{Fun}(\mathcal{A}/\mathcal{G})^*$. The kinematical Hilbert space constructed from a $\text{Diff}_0(X)$-invariant state $\mu \in \text{Fun}(\mathcal{A}/\mathcal{G})$ thus becomes a unitary representation of $\text{Diff}_0(X)$.

It is thus of considerable interest to find a more concrete description of $\text{Diff}_0(X)$-invariant states on the holonomy C*-algebra $\text{Fun}(\mathcal{A}/\mathcal{G})$. In fact, it is not immediately obvious that any exist, in general! For technical reasons, the most progress has been made in the real-analytic case. That is, we take X to be real analytic, $\text{Diff}_0(X)$ to consist of the *real-analytic* diffeomorphisms connected to the identity, and $\text{Fun}(\mathcal{A}/\mathcal{G})$ to be the holonomy C*-algebra generated by real-analytic loops. Here Ashtekar and Lewandowski have constructed a $\text{Diff}_0(X)$-invariant state on $\text{Fun}(\mathcal{A}/\mathcal{G})$ that is closely analogous to the Haar measure on a compact group [4]. They have also given a general characterization of such diffeomorphism-invariant states. The latter was also given by the author [10], using a slightly different formalism, who also constructed many more examples of $\text{Diff}_0(X)$-invariant states on $\text{Fun}(\mathcal{A}/\mathcal{G})$. There is thus some real hope that the loop representation of generally covariant gauge theories can be made rigorous in cases other than the toy models of the previous two sections.

We conclude with some speculative remarks concerning 4d quantum gravity and 2-tangles. The correct inner product on the physical Hilbert space of 4d quantum gravity has long been quite elusive. A path integral formula for the inner product has been investigated recently by Rovelli [48], but there is as yet no manifestly well-defined expression along these lines. On the other hand, an inner product for 'relative states' of quantum gravity in the Kauffman bracket state has been rigorously constructed by the author [9], but there are still many questions about the physics here. The example of 2d Yang–Mills theory would suggest an expression for the inner product of string states

$$\langle \gamma_1, \ldots, \gamma_n | \gamma_1', \ldots, \gamma_n' \rangle$$

as a sum over ambient isotopy classes of surfaces $f \colon \Sigma \to [0,t] \times X$ having the loops γ_i, γ_i' as boundaries. In the case of embeddings, such surfaces are known as '2-tangles', and have been intensively investigated by Carter and Saito [14] using the technique of 'movies'.

The relationships between 2-tangles, string theory, and the loop representation of 4d quantum gravity are tantalizing but still rather obscure. For example, just as the Reidemeister moves relate any two pictures of the same tangle in 3 dimensions, there are a set of movie moves relating any two movies of the same 2-tangle in 4 dimensions. These moves give a set of equations whose solutions would give 2-tangle invariants. For example, the analog of the Yang–Baxter equation is the Zamolodchikov equation, first derived in the context of string theory [26]. These equations can be understood in terms of category theory, since just as tangles form a braided tensor category, 2-tangles form a braided tensor 2-category [25]. It is thus quite significant that Crane [16] has initiated an approach to generally covariant field theory in 4 dimensions using braided tensor 2-categories. This approach also clarifies some of the significance of conformal field theory for 4-dimensional physics, since braided tensor 2-categories can be constructed from certain conformal field theories. In a related development, Cotta-Ramusino and Martellini [15] have endeavored to construct 2-tangle invariants from generally covariant gauge theories, much as tangle invariants may be constructed using Chern–Simons theory. Clearly it will be some time before we are able to appraise the significance of all this work, and the depth of the relationship between string theory and the loop representation of quantum gravity.

Acknowledgements

I would like to thank Abhay Ashtekar, Scott Axelrod, Matthias Blau, Scott Carter, Paolo Cotta-Ramusino, Louis Crane, Jacob Hirbawi, Jerzy Lewandowski, Renate Loll, Maurizio Martellini, Jorge Pullin, Holger Nielsen, and Lee Smolin for useful discussions. Wati Taylor deserves special thanks for explaining his work on Yang-Mills theory to me. Also, I would like to thank the Center for Gravitational Physics and Geometry as a whole for inviting me to speak on this subject.

Bibliography

1.   A. Ashtekar, New variables for classical and quantum gravity, *Phys. Rev. Lett.* **57** (1986) 2244–2247.
      New Hamiltonian formulation of general relativity, *Phys. Rev.* **D36** (1987) 1587–1602.

Lectures on Non-perturbative Canonical Quantum Gravity, Singapore, World Scientific, 1991.

2.   A. Ashtekar, unpublished notes, June 1992.

3.   A. Ashtekar and C. J. Isham, Representations of the holonomy algebra of gravity and non-abelian gauge theories, *Classical & Quantum Gravity* **9** (1992), 1069–1100.

4.   A. Ashtekar and J. Lewandowski, Completeness of Wilson loop functionals on the moduli space of $SL(2, C)$ and $SU(1, 1)$ connections, *Classical & Quantum Gravity* **10** (1993) 673–694.
   Representation theory of analytic holonomy C*-Algebras, this volume.

5.   A. Ashtekar and R. Loll, New loop representations for 2+1 gravity, Syracuse U. preprint.

6.   A. Ashtekar, V. Husain, C. Rovelli, J. Samuel, and L. Smolin, 2+1 gravity as a toy model for the 3+1 theory, *Classical & Quantum Gravity* **6** (1989) L185–L193.

7.   M. Atiyah, *The Geometry and Physics of Knots*, Cambridge U. Press, Cambridge, 1990.

8.   M. Atiyah and R. Bott, The Yang–Mills equations over Riemann surfaces, *Phil. Trans. R. Soc.* **A308** (1983) 523–615.

9.   J. Baez, Quantum gravity and the algebra of tangles, *Classical & Quantum Gravity* **10** (1993) 673–694.

10.   J. Baez, Diffeomorphism-invariant generalized measures on the space of connections modulo gauge transformations, to appear in the proceedings of the Conference on Quantum Topology, eds L. Crane and D. Yetter, hep-th/9305045.
   Link invariants, functional integration, and holonomy algebras, U. C. Riverside preprint, hep-th/9301063.

11.   M. Blau and G. Thompson, Topological gauge theories of antisymmetric tensor fields, *Ann. Phys.* **205** (1991) 130–172.
   Quantum Yang–Mills theory on arbitrary surfaces, *Int. J. Mod. Phys.* **A7** (1992) 3781–3806.

12.   S. Carlip, Six ways to quantize (2+1)-dimensional gravity, U. C. Davis preprint, gr-qc/9305020.

13.   J. S. Carter, *How Surfaces Intersect in Space: an Introduction to Topology*, World Scientific, Singapore, 1993.

14.   J. S. Carter and M. Saito, Reidemeister moves for surface isotopies and their interpretation as moves to movies, U. of South Alabama preprint.
   Knotted surfaces, braid movies, and beyond, this volume.

15.   P. Cotta-Ramusino and M. Martellini, BF theories and 2-knots, this

volume.

16. L. Crane, Topological field theory as the key to quantum gravity, this volume.

17. M. Douglas, Conformal field theory techniques for large N group theory, Rutgers U. preprint, hep-th/9303159.

18. B. Driver, YM$_2$: continuum expectations, lattice convergence, and lassos, *Commun. Math. Phys.* **123** (1989) 575–616.

19. D. Fine, Quantum Yang–Mills on a Riemann surface, *Commun. Math. Phys.* **140** (1991) 321–338.

20. J. Fischer, 2-categories and 2-knots, Yale U. preprint, Feb. 1993.

21. P. Freyd, D. Yetter, J. Hoste, W. Lickorish, K. Millett, and A. Ocneanu, A new polynomial invariant for links, *Bull. Am. Math. Soc.* **12** (1985) 239–246.

22. R. Gambini and A. Trias, Gauge dynamics in the C-representation, *Nucl. Phys.* **B278** (1986) 436–448.

23. J. Gervais and A. Neveu, The quantum dual string wave functional in Yang-Mills theories, *Phys. Lett.* **B80** (1979), 255–258.

24. F. Gliozzi and M. Virasoro, The interaction among dual strings as a manifestation of the gauge group, *Nucl. Phys.* **B164** (1980) 141–151.

25. W. Goldman, The symplectic nature of fundamental groups of surfaces, *Adv. Math.* **54** (1984) 200–225.
Invariant functions on Lie groups and Hamiltonian flows of surface group representations, *Invent. Math.* **83** (1986) 263–302.
Topological components of spaces of representations, *Invent. Math.* **93** (1988) 557–607.

26. D. Gross, Two dimensional QCD as a string theory, U. C. Berkeley preprint, Dec. 1992, hep-th/9212149.

27. D. Gross and W. Taylor IV, Two dimensional QCD is a string theory, U. C. Berkeley preprint, Jan. 1993, hep-th/9301068.
Twists and Wilson loops in the string theory of two dimensional QCD, U. C. Berkeley preprint, Jan. 1993, hep-th/9303046.

28. L. Gross, C. King, and A. Sengupta, Two-dimensional Yang-Mills theory via stochastic differential equations, *Ann. Phys.* **194** (1989) 65–112.

29. G. Horowitz, Exactly soluble diffeomorphism-invariant theories, *Commun. Math. Phys.* **125** (1989) 417–437.

30. A. Jevicki, Loop-space representation and the large-N behavior of the one-plaquette Kogut-Susskind Hamiltonian, *Phys. Rev.* **D22** (1980) 467–471.

31. L. Kauffman, *Knots and Physics,* World Scientific, Singapore, 1991.

32. V. Kazakov, Wilson loop average for an arbitrary contour in two-dimensional $U(N)$ gauge theory, *Nuc. Phys.* **B179** (1981) 283–292.

33. R. Loll, J. Mourão, and J. Tavares, Complexification of gauge theories, Syracuse U. preprint, hep-th/930142.

34. Y. Makeenko and A. Migdal, Quantum chromodynamics as dynamics of loops, *Nucl. Phys.* **B188** (1981) 269–316.
    Loop dynamics: asymptotic freedom and quark confinement, *Sov. J. Nucl. Phys.* **33** (1981) 882–893.

35. D. Marolf, Loop representations for 2+1 gravity on a torus, Syracuse U. preprint, March 1993, gr-qc/9303019.
    An illustration of 2+1 gravity loop transform troubles, Syracuse U. preprint, May 1993, gr-qc/9303019.

36. A. Migdal, Recursion equations in gauge field theories, *Sov. Phys. JETP* **42** (1975) 413–418.

37. J. Minahan, Summing over inequivalent maps in the string theory interpretation of two dimensional QCD, U. of Virginia preprint, hep-th/9301003

38. J. Minahan and A. Polychronakos, Equivalence of two dimensional QCD and the $c = 1$ matrix model, U. of Virginia preprint, hep-th/9305153.

39. S. Naculich, H. Riggs, and H. Schnitzer, Two-dimensional Yang–Mills theories are string theories, Brandeis U. preprint, hep-th/9305097.

40. Y. Nambu, QCD and the string model, *Phys. Lett.* **B80** (1979) 372–376.

41. M. Narasimhan and C. Seshadri, Stable and unitary vector bundles on a compact Riemann surface, *Ann. Math.* **82** (1965) 540–567.

42. M. Nomura, A soluble nonlinear Bose field as a dynamical manifestation of symmetric group characters and Young diagrams, *Phys. Lett.* **A117** (1986) 289–292.

43. P. Peldan, Actions for gravity, with generalizations: a review, U. of Göteborg preprint, May 1993, gr-qc/9305011.

44. V. Periwal, Chern–Simons theory as topological closed string, Institute for Advanced Studies preprint.

45. A. Polyakov, Gauge fields as rings of glue, *Nucl. Phys.* **B164** (1979) 171–188.
    Gauge fields and strings, Harwood Academic Publishers, New York, 1987.

46. J. Pullin, Knot theory and quantum gravity in loop space: a primer, to appear in *Proc. of the Vth Mexican School of Particles and Fields,* ed J. L. Lucio, World Scientific, Singapore; preprint available as

hep-th/9301028.

47. S. Rajeev, Yang–Mills theory on a cylinder, *Phys. Lett.* **B212** (1988) 203–205.

48. C. Rovelli, The basis of the Ponzano–Regge–Turaev–Viro–Ooguri model is the loop representation basis, Pittsburgh U. preprint, April 1993, hep-th/9304164.

49. C. Rovelli and L. Smolin, Loop representation for quantum general relativity, *Nucl. Phys.* **B331** (1990) 80–152.

50. C. Rovelli and L. Smolin, The physical hamiltonian in non-perturbative quantum gravity, Pennsylania State U. preprint, August 1993, gr-qc/9308002.

51. B. Rusakov, Loop averages and partition functions in $U(N)$ gauge theory on two-dimensional manifolds, *Mod. Phys. Lett.* **A5** (1990) 693–703.

52. G. 't Hooft, A two-dimensional model for mesons, *Nucl. Phys.* **B75** (1974) 461–470.

53. E. Witten, 2+1 dimensional gravity as an exactly soluble system, *Nucl. Phys.* **B311** (1988) 46–78.

54. E. Witten, Quantum field theory and the Jones polynomial, *Commun. Math. Phys.* **121** (1989), 351–399.

55. E. Witten, On quantum gauge theories in two dimensions, *Commun. Math. Phys.* **141** (1991) 153–209.
     Localization in gauge theories, Lectures at MIT, February 1992.

56. E. Witten, Chern–Simons gauge theory as a string theory, to appear in the Floer Memorial Volume, Institute for Advanced Studies preprint, 1992.

57. A. Zamolodchikov, Tetrahedron equations and the relativistic *S*-Matrix of straight-strings in 2+1-dimensions, *Commun. Math. Phys.* **79** (1981) 489–505.

58. B. Zwiebach, Closed string field theory: an introduction, MIT preprint, May 1993, hep-th/9305026.

BF Theories and 2-knots

Paolo Cotta-Ramusino

*Dipartimento di Fisica, Universitá di Trento
and INFN, Sezione di Milano,
Via Celoria 16, 20133 Milano, Italy
(email: cotta@milano.infn.it)*

Maurizio Martellini

*Dipartimento di Fisica, Universitá di Milano
and INFN, Sezione di Pavia,
Via Celoria 16, 20133 Milano, Italy
(email: martellini@milano.infn.it)*

Abstract

We discuss the relations between (topological) quantum field theories in 4 dimensions and the theory of 2-knots (embedded 2-spheres in a 4-manifold). The so-called BF theories allow the construction of quantum operators whose trace can be considered as the higher-dimensional generalization of Wilson lines for knots in 3 dimensions. First-order perturbative calculations lead to higher-dimensional linking numbers, and it is possible to establish a heuristic relation between BF theories and Alexander invariants. Functional integration-by-parts techniques allow the recovery of an infinitesimal version of the Zamolodchikov tetrahedron equation, in the form considered by Carter and Saito.

1 Introduction

In a seminal work, Witten [1] has shown that there is a very deep connection between invariants of knots and links in a 3-manifold and the quantum Chern–Simons field theory.

One of the questions that is possible to ask is whether there exists an analog of Witten's ideas in higher dimensions. After all, knots and links are defined in any dimension (as embeddings of k-spheres into a $(k + 2)$-dimensional space), and topological quantum field theories exist in 4 dimensions [2, 3].

The question, unfortunately, is not easy to answer. On one hand, the theory of higher-dimensional knots and links is much less developed; the relevant invariants are, for the time being, more scarce than in the theory of ordinary knots and links. As far as 4-dimensional topological field theories are concerned, only the general BRST structure has been thoroughly discussed; perturbative calculations, at least in the non-Abelian

169

case, do not appear to have been carried out. Even the quantum observables have not been clearly defined (in the non-Abelian case). Moreover, Chern–Simons theory in 3 dimensions benefited greatly from the results of (2-dimensional) conformal field theory. No analogous help is available for 4-dimensional topological field theories.

In this paper we make some preliminary considerations and proposals concerning 2-knots and topological field theories of the BF type. Specifically we propose a set of quantum observables associated to surfaces embedded in 4-space. These observables can be seen as a generalization of ordinary Wilson lines in 3 dimensions. Moreover, the framework in which these observables are defined fits with the picture of 2-knots as 'movies' of ordinary links (or link diagrams) considered by Carter and Saito (see references below). It is also possible to find heuristic arguments that may relate the expectation values of our observables in the BF theory with the Alexander invariants of 2-knots.

As far as perturbative calculations are concerned, they are much more complicated than in the lower-dimensional case. Nevertheless it is possible to recover a relation between first-order calculations and (higher-dimensional) linking numbers, which parallels a similar relation for ordinary links in a 3-dimensional space.

Finally, in 4 dimensions it is also possible to apply functional integration-by-parts techniques. In this case, these techniques produce a solution of an infinitesimal Zamolodchikov tetrahedron equation, consistent with the proposal made by Carter, Saito, and Lawrence.

2 BF theories and their geometrical significance

We start by considering a 4-dimensional Riemannian manifold M, a compact Lie group G, and a principal bundle $P(M, G)$ over M. If A denotes a connection on such a bundle, then its curvature will be denoted by F_A or simply by F. The space of all connections will be denoted by the symbol $\mathcal{A}$ and the group of gauge transformations by the symbol $\mathcal{G}$. The group G, with Lie algebra Lie(G), will be, in most cases, the group $SU(n)$ or $SO(n)$.

Looking for possible topological actions for a 4-dimensional manifold, we may consider [2, 3] the (Gibbs measures of the)

1.   'Chern' action

$$\exp\left(-\kappa \int_M \text{Tr}_\rho(F \wedge F)\right), \tag{2.1}$$

2.   'BF' action

$$\exp\left(-\lambda \int_M \text{Tr}_\rho(B \wedge F)\right). \tag{2.2}$$

In the formulae above κ and λ are coupling constants, ρ is a given representation of the group G and of its Lie algebra. The curvature F is a 2-form on M with values in the associated bundle $P \times_{\text{Ad}} \text{Lie}(G)$ or, equivalently, a tensorial 2-form on P with values in $\text{Lie}(G)$.

The field B in eqn (2.2) is assumed (classically) to be a 2-form of the same nature as F. We shall see, though, that the quantization procedure may force B to take values in the universal enveloping algebra of $\text{Lie}(G)$. It is then convenient to discuss the geometrical aspects of a more general BF theory, where B is assumed to be a 2-form on M with values in the associated bundle $P \times_{\text{Ad}} \mathcal{U}G$ where $\mathcal{U}G$ is the universal enveloping algebra of $\text{Lie}(G)$. In other words B is a section of the bundle

$$\Lambda^2(T^*M) \otimes (P \times_{\text{Ad}} \mathcal{U}G).$$

The space of forms with values in $P \times_{\text{Ad}} \mathcal{U}G$ will be denoted by the symbol $\Omega^*(M, \mathcal{U}\text{ad}P)$: this space naturally contains $\Omega^*(M, \text{ad}P)$, the space of forms with values in $P \times_{\text{Ad}} \text{Lie}(G)$.

In order to obtain an ordinary form on M from an element of $\Omega^*(M, \mathcal{U}\text{ad}P)$, one needs to take the trace with respect to a given representation ρ of $\text{Lie}(G)$. Notice that the wedge product for forms in $\Omega^*(M, \mathcal{U}\text{ad}P)$ is obtained by combining the product in $\mathcal{U}G$ with the exterior product for ordinary forms.

We now mention some examples of B-fields that have a nice geometrical meaning:

1. Let $LM \longrightarrow M$ be the frame bundle and let θ be the soldering form. It is an $\mathbf{R}^n$-valued 1-form, given by the identity map on the tangent space of M. For any given metric g we consider the reduced $SO(n)$-bundle of orthonormal frames $O_g M$. A *vielbein* in physics is defined as the $\mathbf{R}^n$-valued 1-form on M given by $\sigma^*\theta$, for a given section $\sigma: M \longrightarrow O_g M$. We now define the 2-form $B \in \Omega^2(M, \text{ad}O_g M)$ as

$$B \equiv \theta \wedge \theta^T|_{O_g M},$$

where ^{T} denotes transposition. In the corresponding BF action, the curvature F is the curvature of a connection in the orthonormal frame bundle $O_g M$. This action is known to be (classically) equivalent to the Einstein action, written in the vielbein formalism.

2. The manifold $\mathcal{A}$ is an affine space with underlying vector space $\Omega^1(M, \text{ad}P)$. We denote by η the 1-form on $\mathcal{A}$ given by the identity map on the tangent space of $\mathcal{A}$, i.e. on $\Omega^1(M, \text{ad}P)$. From this we can obtain the 2-form on $\mathcal{A}$

$$B \equiv \eta \wedge \eta, \tag{2.3}$$

with values in $\Omega^2(M, \mathcal{U}\text{ad}P)$. More explicitly, B can be defined as

$$B(a,b)(X,Y)|_{p,A} = a(X)b(Y) - a(Y)b(X) - b(X)a(Y) + b(Y)a(X), \tag{2.4}$$

where $a, b \in T_A \mathcal{A};\ X, Y \in T_p P$.

We now want to show that the BF action considered in the last example is connected with the ABJ (Adler–Bell–Jackiw) anomaly in 4 dimensions. On the principal G-bundle

$$P \times \mathcal{A} \longrightarrow M \times \mathcal{A} \tag{2.5}$$

we can consider the tautological connection defined by the following horizontal distribution:

$$\mathrm{Hor}_{p,A} = \mathrm{Hor}_p^A \oplus T_A \mathcal{A} \tag{2.6}$$

where Hor_p^A denotes the horizontal space at $p \in P$ with respect to the connection A. The connection form of the tautological connection is given simply by A, seen as a $(1,0)$-form on $P \times \mathcal{A}$. Its curvature form is given by

$$\mathcal{F}_{p,A} = (F_A)_p + \eta_A. \tag{2.7}$$

Let us now use Q_l to denote an irreducible ad-invariant polynomial on $\mathrm{Lie}(G)$ with l entries. Integrating the Chern–Weil forms corresponding to such polynomials over M we obtain (for $l = 2, 3, 4$)

$$\int_M Q_2(\mathcal{F}, \mathcal{F}) = k_2 \int_M \mathrm{Tr}_\rho(F_A \wedge F_A) \tag{2.8}$$

$$\int_M Q_3(\mathcal{F}, \mathcal{F}, \mathcal{F}) = k_3 \int_M \mathrm{Tr}_\rho(B \wedge F_A) \tag{2.9}$$

$$\int_M Q_4(\mathcal{F}, \mathcal{F}, \mathcal{F}, \mathcal{F}) = k_4 \int_M \mathrm{Tr}_\rho(B \wedge B) \tag{2.10}$$

where B is defined as in eqns (2.3) and (2.4), and k_l are normalization factors. In eqn (2.10) the wedge product also includes the exterior product of forms on $\mathcal{A}$.

As Atiyah and Singer [4] pointed out, one can consider together with

(2.5) the following principal G-bundle[13]:

$$\frac{P \times \mathcal{A}}{\mathcal{G}} \longrightarrow M \times \frac{\mathcal{A}}{\mathcal{G}}. \tag{2.11}$$

Any connection on the bundle $\mathcal{A} \longrightarrow \mathcal{A}/\mathcal{G}$ determines a connection on the bundle (2.11) and hence on the bundle (2.5). When we consider such a connection on (2.5) instead of the tautological one, then the left-hand side of eqn (2.9) becomes the term which generates, via the so-called 'descent equation', the ABJ anomaly in 4 dimensions [4, 5]. Hence the BF action (2.9) is a sort of gauge-fixed version of the generating term for the ABJ anomaly.

In any BF action one has to integrate forms defined on $M \times \mathcal{A}$. Hence we can consider different kinds of symmetries [4, 6]:

1. 'connection invariance': this refers to forms on $M \times \mathcal{A}$ which are closed: when they are integrated over cycles of M, they give closed forms on $\mathcal{A}$.

2. 'gauge invariance': this refer to forms which are defined over $M \times \mathcal{A}$, but are pullbacks of forms defined over $M \times \dfrac{\mathcal{A}}{\mathcal{G}}$.

3. 'diffeomorphism invariance': this can be considered whenever we have an action of $\mathrm{Diff}(M)$ over P and over $\mathcal{A}$. In particular we can consider diffeomorphism invariance when P is a trivial bundle, or when P is the bundle of linear frames over M. In the latter case

$$\frac{P \times \mathcal{A}}{\mathrm{Diff}(M)} \longrightarrow \frac{M \times \mathcal{A}}{\mathrm{Diff}(M)} \tag{2.12}$$

is a principal bundle[14].

4. 'diffeomorphism/gauge invariance': this refers to the action of $\mathrm{Aut}P$, i.e. the full group of automorphisms of the principal bundle P (including the automorphisms which do *not* induce the identity map on the base manifold). This kind of invariance implies gauge invariance and diffeomorphism invariance when the latter is defined.

For instance, the Chern action (2.1) is both connection and diffeomorphism/gauge invariant, while the BF action (2.2) is gauge invariant and its connection invariance depends on further specifications. In particular the BF action of example 2 is connection invariant, while the action of

---

[13]Strictly speaking, in this case, one should consider either the space of irreducible connections and the gauge group, divided by its center, or the space of all connections and the group of gauge transformations which give the identity when restricted to a fixed point of M. We always assume that one of the above choices is made.

[14]In fact one should really consider only the group of diffeomorphisms of M which strongly fix one point of M. Also, instead of $LM \times \mathcal{A}$ it is possible to consider $\mathcal{O}_{M} \times \mathcal{A}^{metric}$, namely the space of all orthonormal frames paired with the *corresponding* metric connections (see e.g. [5]).

example 1 is not. More precisely, in example 1, the integrand of the BF action defines a closed 2-form on $O_g M \times \mathcal{A}$, i.e. we have

$$\delta_{\mathcal{A}}\left(\int_M \mathrm{Tr}_\rho\{F \wedge [\theta \wedge \theta^T]|_{O_g M}\}\right) = 0$$

only when the covariant derivative $d_A \theta$ is zero; in other words, the critical connections are the Levi-Civita connections.

When B is given by (2.3), then for any 2-cycle Σ in M

$$\int_\Sigma \mathrm{Tr}_\rho B = \int_\Sigma Q_2(\mathcal{F} \wedge \mathcal{F})$$

is a closed 2-form on $\mathcal{A}$ and so we have a map

$$\mu : Z_2(M) \longrightarrow Z^2(\mathcal{A}), \tag{2.13}$$

where Z_2 (Z^2) denotes 2-cycles (2-cocycles). When we replace the curvature of the tautological connection with the curvature of a connection in the bundle (2.11), then we again obtain a map (2.13), but now this map descends to a map

$$\mu : H_2(M) \longrightarrow H^2\left(\frac{\mathcal{A}}{\mathcal{G}}\right). \tag{2.14}$$

The map (2.14) is the basis for the construction of the Donaldson polynomials [7].

As a final remark we would like to mention BF theories in arbitrary dimensions. When M is a manifold of dimension d, then we can take the field B to be a $(d-2)$-form, again with values in the bundle $P \times_{\mathrm{Ad}} \mathcal{U}G$. The form (2.4) then makes perfect sense in any dimension.

A special situation occurs when the group G is $SO(d)$. In this case there is a linear isomorphism

$$\Lambda^2(\mathbf{R}^d) \longrightarrow \mathrm{Lie}(SO(d)), \tag{2.15}$$

defined by

$$e_i \wedge e_j \longrightarrow (E_j^i - E_i^j),$$

where $\{e_i\}$ is the canonical basis of $\mathbf{R}^d$ and E_j^i is the matrix with entries $(E_j^i)_n^m = \delta_{m,i}\delta_{n,j}$. The isomorphism (2.15) has the following properties:

1.   It is an isomorphism of inner product spaces, when the inner product in $\mathrm{Lie}(SO(d))$ is given by $(A, B) \equiv (-1/2)\mathrm{Tr}AB$.

2.   The left action of $SO(d)$ on $\mathbf{R}^n$ corresponds to the adjoint action on $\mathrm{Lie}(SO(d))$.

One can then consider a principal $SO(d)$-bundle and an associated bundle E with fiber $\mathbf{R}^d$. In this special case we can consider a different kind of BF theory, where the field B is assumed to be a $(d-2)$-form with values in the $(d-2)$th exterior power of E. In this case the corresponding BF action is given by

$$\exp\left(-\lambda \int_M B \wedge F\right), \qquad (2.16)$$

where the wedge product combines the wedge product of forms on M with the wedge product in $\Lambda^*(\mathbf{R}^d)$. The action (2.16) is invariant under gauge transformations. When the $SO(d)$-bundle is the orthonormal bundle and the field B is the $(d-2)$th exterior power of the soldering form, then the corresponding BF action (2.16) gives the (classical) action for gravity in d dimensions, in the so-called Palatini (first-order) formalism [8, 9].

3   2-knots and their quantum observables

In Witten–Chern–Simons [1] theory we have a topological action on a 3-dimensional manifold M^3, and the observables correspond to knots (or links) in M^3. More precisely to each knot we associate the trace of the holonomy along the knot in a fixed representation of the group G, or 'Wilson line'. In 4 dimensions we have at our disposal the Lagrangians considered at the beginning of the previous section. It is natural to consider as observables, quantities (higher-dimensional Wilson lines) related to 2-knots.

Let us recall here that while an ordinary knot is a 1-sphere embedded in S^3 (or $\mathbf{R}^3$), a 2-knot is a 2-sphere embedded in S^4 or in the 4-space $\mathbf{R}^4$. A generalized 2-knot in a 4-dimensional closed manifold M can be defined as a *closed* surface Σ embedded in M. Two 2-knots (generalized or not) will be called equivalent if they can be mapped into each other by a diffeomorphism connected to the identity of the ambient manifold M.

The theory of 2-knots (and 2-links) is less developed than the theory of ordinary knots and links. For instance, it is not known whether one can have an analogue of the Jones polynomial for 2-knots. On the other hand, one can define Alexander invariants for 2-knots (see e.g. [10]).

The problem we would like to address here is whether there exists a connection between 4-dimensional field theories and (invariants of) 2-knots. Namely, we would like to ask whether there exists a generalization to 4 dimensions of the connection established by Witten between topological field theories and knot invariants in 3 dimensions.

Even though we are not able to show rigorously that a consistent set of non-trivial invariants for 2-knots can be constructed out of 4-dimensional field theories, we can show that there exists a connection between BF theories in 4 dimensions and 2-knots. This connection involves, in different places, the Zamolodchikov tetrahedron equation as well as self-linking and (higher-dimensional) linking numbers.

In Witten's theory one has to perform functional integration upon an observable depending on the given (ordinary) knot (in the given 3-manifold) with respect to a topological Lagrangian. Here we want to do something similar, namely we want to perform functional integration upon different observables depending on a given embedded surface Σ in M with respect to a path integral measure given by the BF action.

The first (classical) observable we want to consider is

$$\mathrm{Tr}_\rho \exp\left(\int_\Sigma \mathrm{Hol}(A, \gamma_{y,x}) B(y) \mathrm{Hol}(A, \gamma_{z,y})\right), \quad y \in \Sigma. \tag{3.1}$$

Here by $\mathrm{Hol}(A, \gamma_{y,x})$ we mean the holonomy with respect to the connection A along the path γ joining two given points y and x belonging to Σ^{15}. Expression (3.1) has the following properties:

1.   It depends on the choice of a map assigning to each $y \in \Sigma$ a path γ joining x and z and passing through y; more precisely, it depends on the holonomies along such paths. One expects that the functional integral over the space of all connections will integrate out this dependency. More precisely, we recall that the (functional) measure over the space of connections is given, up to a Jacobian determinant, by the measure over the paths over which the holonomy is computed. See in this regard the analysis of the Wess–Zumino–Witten model made by Polyakov and Wiegmann [11].

2.   It is gauge invariant if we consider only gauge transformations that are the identity at the points x and z. In the special case when z and x coincide, then (3.1) is gauge invariant without any restriction.

The functional integral of (3.1) can be seen as the vacuum expectation value of the trace (in the representation ρ) of the *quantum surface operator* denoted by $O(\Sigma; x, z)$. In other words we set

$$O(\Sigma; x, z) \equiv \exp\left(\int_\Sigma \mathrm{Hol}(A, \gamma_{y,x}) B(y) \mathrm{Hol}(A, \gamma_{z,y})\right), \quad y \in \Sigma, \tag{3.2}$$

where, in order to avoid a cumbersome notation, we did not write the explicit dependence on the paths γ, and we did not use different symbols for the holonomy and its quantum counterpart (*quantum holonomy*). Moreover, in (3.2) a time-ordering symbol has to be understood.

Note that we will also allow the operator (3.2) to be defined for any embedded surface Σ *closed or with boundary*. In the latter case the points x and z are always supposed to belong to the boundary.

As far as the role of the paths γ considered above is concerned, we recall that, in a 3-dimensional theory with knots, a framing is a (smooth) assignment of a tangent vector to each point of the knot. In 4 dimensions,

---

[15] We assume to have chosen once and for all a reference section for the (trivial) bundle $P|_\Sigma$. Hence the holonomy along a path is defined by the comparison of the horizontally lifted path and the path lifted through the reference section.

instead, we assign a curve to each point of the 2-knot. We will refer to this assignment as a *(higher-dimensional) framing*. We will also speak of a *framed 2-knot*.

4 Gauss constraints

In order to study the quantum theory corresponding to (2.2), we first consider the canonical quantization approach. We take the group G to be $SU(n)$ and consider its fundamental representation; hence we write the field B as $\hat{B} + \sum_a B^a R^a$. Here $\hat{B}$ is a multiple of 1 and $\{R^a\}$ is an orthonormal basis of $\bigoplus_{k=1}^{\infty} UG$. In the BF action we can disregard the $\hat{B}$ component of the field B.

We choose a time direction t and write in local coordinates $(t, x) \equiv (t, x^1, x^2, x^3)$

$$d = d_t + d_x, \quad A = A_0 dt + A_x = A_0 dt + \sum A_i dx^i \quad (i = 1, 2, 3),$$

$$F_A = \sum_i (d_t A_i) \wedge dx^i + (d_x A_0) \wedge dt + \sum_i [A_i, A_0] dx^i \wedge dt + \sum_{i<j} F_{ij} dx^i \wedge dx^j$$

$$B = \sum_{i<j} B_{ij} dx^i \wedge dx^j + B_i dt \wedge dx^i.$$

In local coordinates the action is given by

$$\mathcal{L} \equiv \int_M \mathrm{Tr}(B \wedge F_A) \tag{4.1}$$

$$\approx \int_{M^3 \times I} \sum_{i,j,k} \mathrm{Tr}(B_{ij} d_t A_k + B_i F_{jk} dt + A_0 (DB)_{ijk}) dx^i \wedge dx^j \wedge dx^k,$$

where D denotes the covariant 3-dimensional derivative with respect to the connection A_x.

We first consider the *pure BF* theory, namely, a BF theory where no embedded surface is considered. We perform a Legendre transformation for the Lagrangian $\mathcal{L}$; the conjugate momenta to the fields B_i and A_0, namely $\dfrac{\partial \mathcal{L}}{\partial(\partial_t B_i)}$ and $\dfrac{\partial \mathcal{L}}{\partial(\partial_t A_0)}$, are zero, so we have the primary constraints:

$$\frac{\partial \mathcal{L}}{\partial B_i} = \sum_{j,k} \epsilon_{ijk} (F_{A_x})_{jk} = 0, \quad \frac{\partial \mathcal{L}}{\partial A_0} = \sum_{i,j,k} \epsilon_{ijk} (DB)_{ijk} = 0. \tag{4.2}$$

We do not have secondary constraints since the Hamiltonian is zero; hence the time evolution is trivial. At the formal quantum level the constraints (4.2) will be written as

$$(F_{A_x})_{\mathrm{op}} |\phi_{\mathrm{physical}}\rangle = 0, \quad (DB)_{\mathrm{op}} |\phi_{\mathrm{physical}}\rangle = 0, \tag{4.3}$$

where $_{\rm op}$ stands for 'operator' and the vectors $|\phi_{\rm physical}\rangle$ span the physical Hilbert space of the theory.

We now consider the expectation value of $\mathrm{Tr}_\rho O(\Sigma; x, x)$ (see definition (3.2)). We still want to work in the canonical formalism; the result will be that instead of the constraints (4.2) and (4.3), we will have *a source* represented by the assigned surface Σ. In order to represent the surface operator (3.2) as a source action term to be added to the BF action, we assume that we have a current (singular 2-form) $J_{\rm sing}$, concentrated on the surface Σ, of the form

$$J_{\rm sing} = \sum_a J_{\rm sing}^a R^a.$$

Thus $J_{\rm sing}$ can take values in $\mathcal{U}G$, but has no component in $\mathbf{C} \subset \mathcal{U}(G)$.

We are now in a position to write the operator (3.2) as

$$O(\Sigma; z, z) = \exp\left(\int_M \tilde{\mathrm{Tr}}[\mathrm{Hol}(A, \gamma_{y,z}) B(y) \mathrm{Hol}(A, \gamma_{z,y}) \wedge J_{\rm sing}] \right), \qquad (4.4)$$

where for any $A, B \in \mathcal{U}G$, $A = \sum_a A^a R^a$, $B = \sum_a B^a R^a$, we have set[16]

$$\tilde{\mathrm{Tr}}(AB) = \sum_a A^a B^a.$$

From (4.4) it follows that the component of $J_{\rm sing}$ in $\mathrm{Lie}(G)$ does not give any significant contribution. We now assume $J_{\rm sing} \in \mathcal{U}^2 G$. In this way we neglect possible higher-order terms in $\mathcal{U}^k G \subset \mathcal{U}G$, but this is consistent with our semiclassical treatment.

With the above notation and taking into account the BF action as well, the observable to be functionally integrated is

$$\mathrm{Tr}_\rho \exp\left(\int_M \tilde{\mathrm{Tr}}\{[\mathrm{Hol}(A, \gamma_{y,z}) B(y) \mathrm{Hol}(A, \gamma_{z,y})] \right.$$

$$\left. \wedge [J_{\rm sing} - \mathrm{Hol}(A, \gamma_{z,y})^{-1} F(y) \mathrm{Hol}(A, \gamma_{y,z})^{-1}] \} \right). \qquad (4.5)$$

At this point the introduction of the singular current allows us to represent formally a theory with sources concentrated on surfaces as a pure BF theory with a new 'curvature' given by the difference of the previous curvature and the currents associated to such sources. From eqn (4.4) we conclude, moreover, that $J_{\rm sing}$ as a 2-form should be such that $J_{\rm sing}(y) \wedge d^2\sigma(y) \neq 0$, $y \in \Sigma$, where $d^2\sigma(y)$ is the surface 2-form of Σ.

Now we choose a time direction and proceed to an analysis of the Gauss constraints as in the pure BF theory. We assume that locally the 4-manifold M is given by $M^3 \times I$ (I being a time interval) and the inter-

---

[16] In particular $\tilde{\mathrm{Tr}}$ coincides with Tr when $A, B \in \mathrm{Lie}(G)$.

section of the 3-dimensional manifold $M^3 \times \{t\}$ with the surface Σ is an ordinary link L_t.

In a *neighborhood of* $y \in \Sigma$ the current J_{sing} will look like

$$J^a_{\text{sing}}(x; y) = \delta^{(2)}_\Pi(x - y)R^a \, dx^1_\Pi \wedge dx^2_\Pi, \quad y \in \Sigma, \tag{4.6}$$

where Π is a plane orthogonal to Σ at y (with coordinates x^1_Π and x^2_Π) and $\delta^{(n)}$ denotes the n-dimensional delta function.

The Gauss constraints imply that in the above neighborhood of $y \in \Sigma$ we have

$$F^a(x) = \text{Hol}(A, \gamma_{z,y})J^a_{\text{sing}}(x; y)\text{Hol}(A, \gamma_{y,z}), \ y \in \Sigma, x \in \Pi, \tag{4.7}$$

where γ is a loop in Σ based at z and passing through y. The right-hand side of (4.7) is also equal to

$$\frac{\partial \text{Hol}(A, \gamma_{z,y,z})}{\partial A^a(x)}; \ y \in \Sigma, x \in \Pi.$$

Notice that the previous analysis implies that the components of the curvature F^a (and consequently the components of the connection A^a) must take values in a non-commutative algebra, at least when sources are present. More precisely the components F^a (as well as A^a) should be proportional to $R^a \in \text{Lie}(G)$. Given the fact that B^a and A^a are conjugate fields, we can conclude that the (commutation relations of the) quantum theory will require B^a to be proportional to R^a as well.

As in the lower-dimensional case (ordinary knots (or links) as sources in a 3-dimensional theory, [12]) the restriction of the connection A to the plane Π looks (in the approximation for which $\text{Hol}(A, \gamma_{z,x,z}) = 1$) like

$$A^a(x)|_\Pi = R^a \, d\log(x - y) \tag{4.8}$$

for a given $y \in \Sigma$ (here d is the exterior derivative).

In order to have some geometrical insight into the above connection, we consider a linear approximation, where the surface Σ can be approximated by a collection of 2-dimensional planes in $\mathbf{R}^4$, i.e. by a collection of hyperplanes in $\mathbf{C}^2$. These hyperplanes, and the hyperplane Π considered above, are assumed to be in general position. This means that the intersection of the hyperplane Π with any other hyperplane is given by a point.

The quantum connection (4.8) on Π gives a representation of the first homotopy group of the manifold of the arrangements of hyperplanes (points) in Π, i.e. more simply, of the (pure) braid group [1, 13]. It is then natural to ask whether the same is true in 4 dimensions, namely whether the 4-dimensional connection, corresponding to the (critical) curvature (4.7) in the linear approximation defined above), is related to the higher braid group [14]. We will discuss this point further elsewhere; let us mention only that the 4-dimensional critical connection is related to the existence of a higher-dimensional version of the Knizhnik–Zamolodchikov equation associated with the representation of the higher braid groups, as suggested by Kohno [15].

5   Path integrals and the Alexander invariant of a 2-knot

In the previous section we worked specifically in the canonical approach. When instead we consider a covariant approach, and compute the expectation values, with respect to the BF functional measure only, we can write[17]

$$\langle \mathrm{Tr}_\rho \, O(\Sigma; x, x) \rangle =$$

$$\int \mathcal{D}A\mathcal{D}B \, \mathrm{Tr}_\rho \, \exp\left(\tilde{\mathrm{Tr}} \int_M B(x) \wedge (F(x) - \mathrm{Hol}(A, \gamma_{x,z}) J_{\mathrm{sing}}(x) \mathrm{Hol}(A, \gamma_{z,x})) \right) \tag{5.1}$$

Again, in the approximation in which $\mathrm{Hol}(A, \gamma_{x,z}) = 1$, we can set

$$\langle \mathrm{Tr}_\rho \, O(\Sigma; x, x) \rangle \approx \int \mathcal{D}A\mathcal{D}B \, \mathrm{Tr}_\rho \, \exp\left(\tilde{\mathrm{Tr}} \int_M B(x) \wedge (F(x) - J_{\mathrm{sing}}(x)) \right). \tag{5.2}$$

A rough (and formal) estimate of the previous expectation value (5.2) can be given as follows.

By integration over the B field we obtain something like

$$\langle \mathrm{Tr}_\rho \, O(\Sigma; x, x) \rangle \approx \dim(\rho) \int \mathcal{D}(A) \delta(F - J_{\mathrm{sing}}), \tag{5.3}$$

using a functional Dirac delta. By applying the standard formula for Dirac deltas evaluated on composite functions, we heuristically obtain

$$\langle \mathrm{Tr}_\rho O(\Sigma; x, x) \rangle \approx \dim(\rho) \, |\det(D_{A_0}|_{M \backslash \Sigma})|^{-1}, \tag{5.4}$$

where $\det(D_{A_0}|_{M \backslash \Sigma})$ denotes the (regularized) determinant of the covariant derivative operator with respect to a background *flat* connection A_0 on the space $M \backslash \Sigma$.

It is then natural to interpret the expectation value (5.4) (with the Faddeev–Popov ghosts included) as the analytic (Ray–Singer) torsion for the complement of the 2-knot Σ (see [16, 17]). This torsion is related to the Alexander invariant of the 2-knot [18].

---

[17]We omit the ghosts and gauge-fixing terms [3], since they are not relevant for a rough calculation of (5.1).

6 First-order perturbative calculations and higher-dimensional linking numbers

Let us now consider the expectation value of the quantum surface observable $\text{Tr}_\rho\, O(\Sigma; x, x)$, *with respect to the total BF + Chern action.* The two fields involved in the quantum surface operator are A_μ^a (the connection) and $B_{\mu,\nu}^a$ (the B-field) with $a = 1, \ldots, \dim(G)$ and $\mu, \nu = 1, \ldots, 4$. The connection A is present in the quantum surface operator via the holonomy of the paths $\gamma_{y,x}$ and $\gamma_{x,y}$ ($y \in \Sigma$) as in definitions (3.1) and (3.2).

At the first-order approximation in perturbation theory (with a background field and covariant gauge), we can write the holonomy as

$$\text{Hol}(A, \gamma_{y,x}) \approx 1 + \kappa \int_{\gamma_{y,x}} \sum_\mu A_\mu(z)dz^\mu + \cdots.$$

At the same order, the only relevant part of the BF action is the kinetic part $(B \wedge dA)$, so we get the following Feynman propagator:

$$\langle A_\mu^a(x) B_{\nu\rho}^b(y) \rangle = \frac{2i}{\lambda} \delta^{ab} \frac{1}{4\pi^2} \sum_\tau \epsilon_{\mu\nu\rho\tau} \frac{(x^\tau - y^\tau)}{|x - y|^4}. \tag{6.1}$$

In order to avoid singularities, we perform the usual point-splitting regularization. This is tantamount to lifting the loops $\gamma_{x,y,x}$ in a neighborhood of y. The lifting is done in a direction which is normal to the surface Σ.

The complication here is that the loop γ itself (based at x and passing through y) depends on the point y. While the general case seems difficult to handle, we can make simplifying choices which appear to be legitimate from the point of view of the general framework of quantum field theory. For instance, we can easily assume that when we assign to a point y the loop $\gamma_{x,y,x}$ based at x and passing through y, then the same loop will be assigned to any other point y' belonging to γ.

Moreover, we can consider an arbitrarily fine triangulation T of Σ and take into consideration only one loop γ_T which has non-empty intersection with each triangle of T. This approximation will break gauge invariance, which will be, in principle, recovered only in the limit when the size of the triangles goes to zero. (For a related approach see [19].) The advantage of this particular approximation is that we have only to deal with one loop γ_T, which by point-splitting regularization is then completely lifted from the surface. We call this lifted path γ_T^r, where r stands for regularized.

Finally we get (to first-order approximation in perturbation theory and up to the normalization factor $\langle 1 \rangle^{18}$)

$$\langle \mathrm{Tr}_\rho\, O(\Sigma; x, x) \rangle \approx$$

$$\dim(\rho)\left[1 + C_2(\rho)\frac{2i\kappa}{\lambda}\frac{1}{4\pi^2}\sum_{\mu,\nu,\rho,\tau}\epsilon_{\mu\nu\rho\tau}\int_\Sigma dx^\mu dx^\nu \int_{\gamma_T^r} dy^\rho \frac{(x^\tau - y^\tau)}{|x-y|^4} + \cdots\right],$$

$$(6.2)$$

where $C_2(\rho)$ is the trace of the quadratic Casimir operator in the representation ρ. When Σ is a 2-sphere, then (6.2) is given by

$$\dim(\rho)\left[1 + C_2(\rho)\frac{2i\kappa}{\lambda}\mathcal{L}(\gamma_T^r, \Sigma) + \cdots\right]$$

where $\mathcal{L}(\gamma_T^r, \Sigma)$ is the (higher-dimensional) *linking number* of Σ with γ_T^r [20].

7   Wilson 'channels'

Let Σ again represent our 2-knot (surface embedded in the 4-manifold M). Carter and Saito [21, 22], inspired by a previous work by Roseman [23], describe a 2-knot Σ (in fact an embedded 2-sphere) in 4-space by the 3-dimensional analogue of a knot diagram: they project down Σ to a 3-dimensional space, keeping track of the over- and under-crossings. One of the coordinates of this 3-space is interpreted as 'time'. The intersection of this 3-dimensional diagram of a 2-knot with a plane (at a fixed time) gives an ordinary link diagram, while the collection of all these link diagrams at different times ('stills') gives a 'movie' representing the 2-knot.

In the Feynman formulation of field theory, we start by considering the surface Σ embedded in the 4-manifold M directly, not its projection. At the local level M will be given by $M^3 \times I$ (I being a time interval), so that the intersection of the 3-dimensional manifold $M^3 \times \{t\}$ with Σ gives an ordinary link L_t (for all times t) in M^3. In order to connect quantum field theory with the standard theory of 2-knots, we assume more simply that M is the 4-space $\mathbf{R}^4$ and that Σ is a 2-sphere. One coordinate axis in $\mathbf{R}^4$ is interpreted as time, and the intersection of Σ with 3-space (at fixed times) is given by ordinary links, with the possible exception of some critical values of the time parameter.

In other words, we have a space projection $\pi_t\colon \mathbf{R}^4 \longrightarrow \mathbf{R}^3$ for each time t, so that

$$L_t(\Sigma) \equiv \pi_t^{-1}(\mathbf{R}^3) \cap \Sigma$$

is, for non-critical times, an ordinary link. For any time interval (t_1, t_2) we

---

[18] While the normalization factor for a pure (4-dimensional) BF theory appears to be trivial [17], in a theory with a combined BF + Chern action we have a contribution coming from the Chern action. This contribution has been related to the first Donaldson invariant of M [2].

will also use the notation $\Sigma_{(t_1,t_2)} \equiv \bigcup_{t \in (t_1,t_2)} \{\pi_t^{-1}(\mathbf{R}^3) \cap \Sigma\}$.

We consider a set $\mathcal{T}$ of (non-critical) times $\{t_i\}_{i=1,\cdots,n}$. We denote the components of the corresponding links L_{t_i} by the symbols $K_i^{j(i)}$, $j = 1,\ldots,s(i)$. Here and below we will write $j(i)$ simply to denote the fact that the index j ranges over a set depending on i. We choose one base-point $x_i^{j(i)}$ in each knot $K_i^{j(i)}$ and we denote by $\Sigma_i^{j(i),j(i+1)}$ the surface contained in Σ whose boundary includes $K_i^{j(i)}$ and $K_{i+1}^{j(i+1)}$. We then assign a framing to $\Sigma_i^{j(i),j(i+1)}$ as follows: for each point p in the interior of $\Sigma_i^{j(i),j(i+1)}$ we associate a path with endpoints $x_i^{j(i)}$ and $x_{i+1}^{j(i+1)}$ passing through p. To the surface $\Sigma_i^{j(i),j(i+1)}$ *with boundary components* $K_i^{j(i)}$ and $K_{i+1}^{j(i+1)}$, we associate the operator

$$\mathrm{Hol}(A, K_i^{j(i)})_{x_i^{j(i)}} \, O(\Sigma_i^{j(i),j(i+1)}; x_i^{j(i)}, x_{i+1}^{j(i+1)}) \, \mathrm{Hol}(A, K_{i+1}^{j(i+1)})_{x_{i+1}^{j(i+1)}},$$

$$(7.1)$$

where the subscript to the term $\mathrm{Hol}(A,\ldots)$ denotes the basepoint.

We refer to the 2-knot Σ equipped with the set $\mathcal{T}$ of times as the *(temporally) sliced 2-knot*. For the sliced 2-knot, we define the *Wilson channels* to be the operators of the form

$$O(\Sigma_{(-\infty,t_1)}) \, \mathrm{Hol}(A, K_1^1)_{x_1^1} \cdots$$

$$\mathrm{Hol}(A, K_i^{j(i)})_{x_i^{j(i)}} O(\Sigma_i^{j(i),j(i+1)}; x_i^{j(i)}, x_{i+1}^{j(i+1)}) \, \mathrm{Hol}(A, K_{i+1}^{j(i+1)})_{x_{i+1}^{j(i+1)}}$$

$$\cdots O(\Sigma_{(t_n,+\infty)}; x_n^{j(n)}, +\infty). \qquad (7.2)$$

We assume the following requirements for the Wilson channels:

1. The operators which appear in a Wilson channel are chosen in such a way that a surface operator is sandwiched between knot operators only if the relevant knots belong to the boundary of the surface.

2. The paths that are included in the surface operators $O(\Sigma_i^{j(i),j(i+1)})$ are not allowed to touch the boundaries of the surface, except at the initial and final point.

Finally, to the 'sliced' 2-knot we associate the product of the traces of all possible Wilson channels. This product can be seen as a possible higher-dimensional generalization of the Wilson operator for ordinary links in 3-space that was considered by Witten.

We would like now to compare the Wilson channel operators associated to a sliced 2-knot Σ with the operator

$$O(\Sigma; -\infty, +\infty) \qquad (7.3)$$

considered in Section 5 (no temporal slicing involved). The main difference concerns the prescriptions that are given involving the paths γ that enter the definition of (7.3) (or, equivalently, the framing of the 2-knot). The Wilson channels are obtained from (7.3) by:

1. choosing a finite set of times $\mathcal{T} \equiv \{t_i\}$, and
2. requiring the paths γ to *follow one of the components of the link* L_t, for each $t \in \mathcal{T}$.

Finally, recall that in the Wilson channel operators, each of the paths γ is forbidden to follow two separate components of the same link L_t, for any time $t \in \mathcal{T}$. This last condition can be understood by noticing that in order to follow both the jth and the j'th components of the link L_{t_i}, for some $i, j(i), j'(i)$, one path should follow the jth component, then enter the surface $\Sigma_i^{j(i),j(i+1)}$ (which is assumed to be equal to $\Sigma_i^{j'(i),j''(i+1)}$), and finally go *back* to the j'th component, thus contradicting a (local) causality condition.

8 Integration by parts

The results of the previous section imply that, given a temporally sliced 2-knot Σ, we can construct quantum operators by considering all the Wilson channels relevant to surfaces $\Sigma_i^{j(i),j(i+1)}$, $i \leq l$ ($i \geq l$), and links L_{t_i} for $i < l$ ($i > l$). Let us denote these quantum operators by the terms

$$W^{\text{in}}(\Sigma, t_l), \quad W^{\text{out}}(\Sigma, t_l). \tag{8.1}$$

They can be seen as a sort of convolution, depending on Σ, of all the quantum holonomies associated to the links L_{t_i} for $i < l$ ($i > l$).

Ordinary links are the closure of braids, and one may describe the surface which generates the different links at times $\{t_i\}$ as a sequence (movie) of braids [24]. But a link is also obtained by closing a tangle, and the time evolution of a link (represented by a surface) can also be described in terms of 2-tangles, i.e. movies of (ordinary) tangles. It has been shown that 2-tangles form a (rigid, braided) 2-category [25]. The 2-tangle approach (with its 2-categorical content) may reveal itself as a useful one for quantum field theory.

The initial tangle is called the source and the final tangle is called the target. In our case, say, the source represents L_{t_l} while the target represents $L_{t_{l'}}$, for $l' > l$. The quantum counterpart of the source and target is represented by the quantum holonomies associated to the corresponding links, while the quantum counterpart of the 2-tangle is represented by the quantum surface operator relevant to the surface $\Sigma_{(t_l, t_{l'})}$.

One can think of the quantum surface operators as acting on quantum holonomies (by convolution), but it is difficult to do *finite* calculations and write this action explicitly. What we can do instead is to make some *infinitesimal* calculations using integration-by-parts techniques in Feynman path integrals.

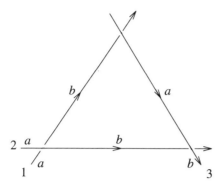

FIG. 1. Three strings

Generally speaking, in a finite time evolution $t_l \longrightarrow t_{l'}$, the quantum surface operator (defined by the surface with boundaries L_{t_l} and $L_{t_{l'}}$) will map the operator $W^{\text{in}}(\Sigma, t_l)$ into the operator $W^{\text{in}}(\Sigma, t_{l'})$. Let us now consider what can be seen as the 'derivative' of this map. Namely, let us see what happens when we consider the quantum operator corresponding to a surface bordering two links L_{t_l} and L'_{t_l} that differ by an elementary change. *What the functional integration-by-parts rules will show, is that one can interchange a variation of the surface operator with a variation of the link (and consequently of its quantum holonomy).* So in order to compute the 'derivative' of the surface operator, we may consider variations of the link L_{t_l}.

In order to describe elementary changes of links, recall that braids or tangles (representing links at a fixed time) can be seen as collections of oriented strings, while 2-tangles describe the interaction of those strings [26]. We consider now two links that differ from each other only in a (small) region involving three strings. Let us number these strings with numbers $1, 2, 3$. The string 1 crosses under the string 2, then the string 2 crosses under the string 3, and the string 1 crosses under the string 3. We denote by the symbols a and b the part of each string that precedes and, respectively, follows the first crossing, as shown in Fig. 1.

The integration-by-parts rules for the BF theory are as follows. We consider a knot $K_l^{j(l)}$, which we denote here by C in order to simplify the notation, and a point $x \in C$. By the symbol δ_x we mean the variation (of C) obtained by inserting an infinitesimally small loop c_x in C at the point x. We obtain

$$\delta_x \left(\text{Hol}(A, C)_x \right) \exp\left(-\lambda \int_M \text{Tr}(B \wedge F) \right)$$

$$= \sum_{\mu,\nu,a} d^2 c_x^{\mu\nu} \text{Hol}(A, C)_x \, F^a_{\mu\nu} \, R^a \, \exp\left(-\lambda \int_M \text{Tr}(B \wedge F) \right)$$

$$= \frac{-1}{2\lambda} \sum_{\mu,\nu,\rho,\tau,a} \epsilon_{\mu\nu\rho\tau} \, d^2 c_x^{\mu\nu} \, \exp\left(-\lambda \int_M \mathrm{Tr}(B\wedge F)\right) \mathrm{Hol}(A,C)_x \, R^a \, \frac{\delta}{\delta B_{\rho\tau}^a(x)},$$

$$(8.2)$$

where $d^2 c_x$ denotes the surface 2-form of the surface bounded by the infinitesimal loop c_x, R^a denotes as usual an orthonormal basis of Lie$(SU(n))$, and the trace is meant to be taken in the fundamental representation. The second equality has been obtained by *functional integration by parts*; in this way we produced the functional derivative with respect to the B-field. The integration-by-parts rules of the BF theory in 4 dimensions completely parallel the integration-by-parts rules of the Chern–Simons theory in 3 dimensions [27]. When we apply the functional derivative $\dfrac{\delta}{\delta B_{\rho\tau}^a(x)}$ to a surface operator $O(\Sigma'; x', x'')$ for a surface Σ' whose boundary includes C^{19} we obtain

$$\frac{\delta}{\delta B_{\rho\tau}^a(x)} O(\Sigma'; x', x'') = R^a \delta^{(4)}(x - x') d^2 \sigma'^{\rho,\tau} O(\Sigma'; x', x''), \qquad (8.3)$$

where $d^2 \sigma'$ is the surface 2-form of Σ'.

From eqn (8.2) we conclude that only the variations c_x for which $d^2\sigma' \wedge d^2 c_x \neq 0$ matter in the functional integral. But these variations are exactly the ones for which the small loop c_x is such that the surface element $d^2 c_x$ is transverse to the knot C (as in [27]).

Now we come back to the link L_{t_l} with the three crossings as in Fig. 1. In order to consider the more general situation we assume that the three strings in Fig. 1 belong to different components K_l^1, K_l^2, K_l^3 of L_{t_l} that will evolve into three different components K_{l+1}^1, K_{l+1}^2, K_{l+1}^3 of $L_{t_{l+1}}$. By performing three such small variations at each one of the crossings of the three strings considered in Fig. 1, we can switch each under-crossing into an over-crossing. Now as in [27] we compare the expectation value of the original configuration with the configuration with the three crossings switched. Such comparison, with the aid of eqns (8.2) and (8.3) and the Fierz identity [27], leads to the conclusion[20] that when we consider the action of the surface operators $O(\Sigma_l^{j,j}; x_l^j, x_{l+1}^j)$, $j = 1, 2, 3$, on the quantum holonomies of the knots K_l^j, its infinitesimal variation δO can be expressed as follows:

$$(\delta O)|\cdots W_l \cdots |0\rangle = \mathcal{R}|\cdots W_l \cdots |0\rangle. \qquad (8.4)$$

Here W_l is a short notation for the operator

---

[19]We are considering here only the small variations of Σ', so we can disregard the holonomies of the paths γ in the surface operators; the only relevant contribution to be considered is the one given by the field $B = \sum B^a R^a$.

[20]Details will be discussed elsewhere.

$$W_l \equiv \mathrm{Hol}(A, K_l^1)_{x_l^1} O(\Sigma_l^{1,1}; x_l^1, x_{l+1}^1) \mathrm{Hol}(A, K_{l+1}^1)_{x_{l+1}^1}$$

$$\otimes \mathrm{Hol}(A, K_l^2)_{x_l^2} O(\Sigma_l^{2,2}; x_l^2, x_{l+1}^2) \mathrm{Hol}(A, K_{l+1}^2)_{x_{l+1}^2}$$

$$\otimes \mathrm{Hol}(A, K_l^3)_{x_l^3} O(\Sigma_l^{3,3}; x_l^3, x_{l+1}^3) \mathrm{Hol}(A, K_{l+1}^3)_{x_{l+1}^3};$$

the dots in (8.4) involve the operators $W^{\mathrm{in}}(\Sigma, t_l)$ and $W^{\mathrm{out}}(\Sigma, t_l)$ (see (8.1)), the vector space associated to the string j is given by a tensor product $V_a^j \otimes V_b^j$, and finally $\mathcal{R}$ is a suitable representation of the following operator:

$$\left(1 + (1/\lambda) \sum_a R^a \otimes R^a + \cdots\right)_{1a,2a} \left(1 + (1/\lambda) \sum_a R^a \otimes R^a + \cdots\right)_{1b,3a}$$

$$\left(1 + (1/\lambda) \sum_a R^a \otimes R^a + \cdots\right)_{2b,3b}.$$

Now we recall that $1 + (1/\lambda) \sum_a R^a \otimes R^a$ is the infinitesimal approximation of a quantum Yang–Baxter matrix R. Hence the calculations of this section suggest that, at the finite level, the solution of the Zamolodchikov tetrahedron equation which may be more relevant to topological quantum field theories is the one considered in [28, 29], namely

$$S_{123} = R_{1a,2a} R_{1b,3a} R_{2b,3b},$$

where R is a solution of the quantum Yang–Baxter equation.

Acknowledgements

S. Carter and M. Saito have kindly explained to us many aspects of their work on 2-knots, both in person and via email. We thank them very much. We also thank A. Ashtekar, J. Baez, R. Capovilla, L. Kauffman, and J. Stasheff very much for discussions and comments. Both of us would also like to thank J. Baez for having organized a very interesting conference and for having invited us to participate. P. C.-R. would like to thank S. Carter and R. Capovilla for invitations to the University of South Alabama and Cinvestav (Mexico), respectively.

Bibliography

1. E. Witten, Quantum field theory and the Jones polynomial, Commun. Math. Phys. **121** (1989) 351–399.

2. E. Witten, Topological quantum field theory, Commun. Math. Phys. **117** (1988) 353–386.

3. D. Birmingham, M. Blau, M. Rakowski, and G. Thompson, Topological field theory, Phys. Rep. **209** (1991) 129–340.

4. M. F. Atiyah and I. M. Singer, Dirac operators coupled to vector

potentials, Proc. Natl. Acad. Sci. **81** (1984) 2597–2600.

5. L. Bonora, P. Cotta-Ramusino, M. Rinaldi, and J. Stasheff, The evaluation map in field theory, sigma-models and strings, Commun. Math. Phys. **112** (1987) 237–282 and **114** (1988) 381–437.

6. L. Baulieu and I. M. Singer, Topological Yang–Mills symmetry, Nucl. Phys. (Proc. Suppl.) **B5** (1988) 12–19.

7. S. K. Donaldson and P. K. Kronheimer, *The Geometry of Four-Manifolds*, Oxford University Press (1990).

8. R. Capovilla, T. Jacobson, and J. Dell, Gravitational instantons as $SU(2)$ gauge fields, Classical & Quantum Gravity **7** (1990) L1–L3.

9. A. Ashtekar, C. Rovelli, and L. Smolin, Weaving a classical metric with quantum threads, Phys. Rev. Lett. **69** (1992) 237–240.

10. D. Rolfsen, *Knots and Links*, Publish or Perish, Inc., Wilmington, Delaware (1976).

11. A. M. Polyakov and P. B. Wiegmann, Goldstone fields in two dimensions with multivalued actions, Phys. Lett. **B141** (1984) 223–228.

12. E. Guadagnini, M. Martellini, and M. Mintchev, Braids and quantum group symmetry in Chern–Simons theory, Nucl. Phys. **B336** (1990) 581–609.

13. T. Kohno, Monodromy representation of braid groups and Yang–Baxter equations, Ann. Inst. Fourier **37** (1987) 139–160.

14. Yu. Manin and V. V. Schechtman, Arrangements of hyperplanes, higher braid groups and higher Bruhat orders, Adv. Studies Pure Math. **17** (1989) 289–308.

15. T. Kohno, Integrable connections related to Manin and Schechtman's higher braid groups, preprint.

16. A. S. Schwarz, The partition function of degenerate quadratic functionals and Ray–Singer invariants, Lett. Math. Phys. **2** (1978) 247–252.

17. M. Blau and G. Thompson, A new class of topological field theories and the Ray–Singer torsion, Phys. Lett. **B228** (1989) 64–68.

18. V. G. Turaev, Reidemeister torsion in knot theory, Russ. Math. Surv. **41** (1986) 119–182.

19. I. Ya. Aref'eva, Non Abelian Stokes formula, Teor. Mat. Fiz. **43** (1980) 111–116.

20. G. T. Horowitz and M. Srednicki, A quantum field theoretic description of linking numbers and their generalization, Commun. Math. Phys. **130** (1990) 83–94.

21. J. S. Carter and M. Saito, Syzygies among elementary string interactions of surfaces in dimension 2+1, Lett. Math. Phys. **23** (1991) 287–300.

22. J. S. Carter and M. Saito, Reidemeister moves for surface isotopies and their interpretation as moves to movies, *J. Knot Theory Ramifications*, **2** (1993) 251–284.

23. D. Roseman, Reidemeister-type moves for surfaces in four-dimensional space, preprint (1992).

24. J. S. Carter and M. Saito, Knotted surfaces, braid movies, and beyond, this volume.

25. J. E. Fischer, Jr, Geometry of 2-categories, PhD Thesis, Yale University (1993).

26. A. B. Zamolodchikov, Tetrahedra equations and integrable systems in 3-dimensional space, Sov. Phys. JETP **52** (1980) 325–336.

27. P. Cotta-Ramusino, E. Guadagnini, M. Martellini, and M. Mintchev, Quantum field theory and link invariants, Nucl. Phys. **B330** (1990) 557–574.

28. J. S. Carter and M. Saito, On formulations and solutions of simplex equations, preprint (1992).

29. R. J. Lawrence, Algebras and triangle relations, preprint (1992).

Knotted Surfaces, Braid Movies, and Beyond

J. Scott Carter

Department of Mathematics, University of South Alabama,
Mobile, Alabama 36688, USA
(email: carter@mathstat.usouthal.edu)

Masahico Saito

Department of Mathematics, University of Texas at Austin,
Austin, Texas 78712, USA
(email: saito@math.utexas.edu)

Abstract

Knotted surfaces embedded in 4-space can be visualized by means of a variety of diagrammatic methods. We review recent developments in this direction. In particular, generalizations of braids are explained. We discuss generalizations of the Yang–Baxter equation that correspond to diagrams appearing in the study of knotted surfaces. Our diagrammatic methods of producing new solutions to these generalizations are reviewed. We observe 2-categorical aspects of these diagrammatic methods. Finally, we present new observations on algebraic/algebraic-topological aspects of these approaches. In particular, they are related to identities among relations that appear in combinatorial group theory, and cycles in Cayley complexes. We propose generalizations of these concepts from braid groups to groups that have certain types of Wirtinger presentations.

1 Introduction

It has been hoped that ideas from physics can be used to generalize quantum invariants to higher-dimensional knots and manifolds. Such possibilities were discussed in this conference.

In dimension 3 Jones-type invariants were defined algebraically via braid group representations or diagrammatically via Kauffman brackets. The first step, then, in the 1-dimensionally higher case is to get algebraic or diagrammatic ways to describe knotted surfaces embedded in 4-space.

In this paper, we first review recent developments in this direction. Reidemeister-type moves were generalized by Roseman [38] and their movie versions were classified by the authors [4, 7]. On the other hand, various braid forms for surfaces were considered by several other authors. In particular, Kamada [26] proved generalized Alexander's and Markov's theorems for surface braids defined by Viro. We combined the movie and the surface braid approaches to obtain moves for the braid movie form of knotted surfaces. The braid movie description is rich in algebraic and diagrammatic flavor.

The exposition in Section 2 provides the basic topological tools for finding new invariants along these lines if they exist.

On the other hand, knot theoretical, diagrammatic techniques have been used to solve algebraic problems related to statistical physics. In particular, we gave solutions to various types of generalizations of the quantum Yang–Baxter equation (QYBE). Such generalizations originated in the work of Zamolodchikov [43]. In Section 3 we review our diagrammatic methods in solving these equations.

Such generalizations appeared in the context of 2-categories [27] and higher algebraic structures [33]. In Section 4 we follow Fischer [14] to show that braid movies form a braided monoidal 2-category in a natural way.

New observations in this paper are in Section 5, where we discuss close relations between braid movie moves and various concepts in algebra and algebraic topology. Some of the moves correspond to 2-cycles in Cayley complexes. Others are related to Peiffer equivalences that are studied in combinatorial group theory and homotopy theory. The charts defined in Section 2 can be regarded as 'pictures' (that are defined for any group presentations) of null homotopy disks in classifying spaces. We propose generalized chart moves for groups that have Wirtinger presentations.

2   Movies, surface braids, charts, and isotopies

2.1   Movie moves

The motion picture method for studying knotted surfaces is one of the most well known. It originated with Fox [15] and has been developed and used extensively (for example [15, 23, 29, 30, 36]). In [4, 7] we gave a set of Reidemeister moves for movies of knotted surfaces such that two movies described isotopic knottings if and only if one could be obtained from the other by means of certain local changes in the movies. In order to depict the movie moves, we first projected a knot onto a 3-dimensional subspace of 4-space.

Here a map from a surface to 3-space is called *generic* if every point has a neighborhood in which the surface is either (1) embedded, (2) two planes intersecting along a double-point curve, (3) three planes intersecting at an isolated triple point, or (4) the cone on the figure eight (also called Whitney's umbrella). Such a map is called generic because the collection of maps of this form is an open dense subset of the space of all the smooth maps; see [18]. Local pictures of these are depicted in Fig. 1.

Then we fix a height function on the 3-space. By cutting 3-space by level planes, we get immersed circles (with finitely many exceptions where critical points occur). We also can include crossing information by breaking circles at underpasses. By a *movie* we mean a sequence of such curves

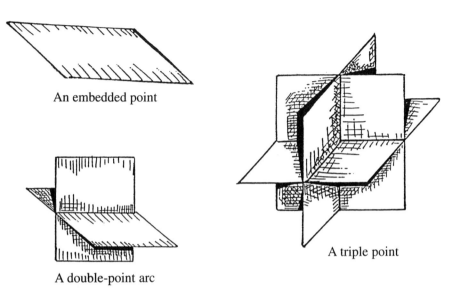

An embedded point

A double-point arc

A triple point

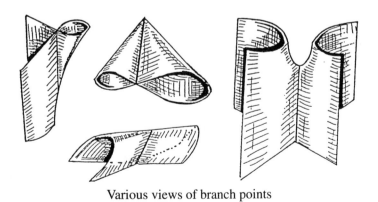

Various views of branch points

FIG. 1. Generic intersection points

on the planes. Each element in a sequence is called a *still*. Again without loss of generality we can assume that successive stills differ by at most a classical Reidemeister move or a critical point of the surface. The Reidemeister moves are interpreted as critical points of the double points of the projected surface, and so the movies contain critical information for the surface and for the projection of the surface. Finally, the moves to these movies were classified following Roseman [38] by means of an analysis of fold-type singularities and intersection sets.

The movie moves are depicted in Figs 3 and 4. In these figures each still represents a local picture in a larger knot diagram. The different stills differ by either a Reidemeister move or critical point (birth/death or saddle) on the surface. These are called the *elementary string interactions (ESIs)*. Figure 2 depicts ESIs. The moves indicate when two different sequences of ESIs depict the same knotting. Any two movie descriptions of the same knot are related by a sequence of these moves. One rather large class of moves is not explicitly given in this list: namely, distant ESIs commute; that is, when two string interactions occur on widely separated segments, the order of the interactions can be switched.

2.2   Surface braids

In April 1992, Oleg Viro described to us the braided form of a knotted surface. Similar notions had been considered by Lee Rudolph [39] and X. S. Lin [35]. Kamada [22-26] has studied Viro's version of surface braids; we follow Kamada's descriptions.

2.2.1   *Definitions*

Let B^2 and D^2 denote 2-disks. A *surface braid* is a compact oriented surface F properly embedded in $B^2 \times D^2$ such that

(1) the projection $p: B^2 \times D^2 \to D^2$ to the second factor restricted to F, $p|_F: F \to D^2$, is a branched covering;

(2) The boundary of F is the standard unlink: $\partial F = (n \text{ points}) \times \partial D^2 \subset B^2 \times \partial D^2 \subset \partial(B^2 \times D^2)$ where n points are fixed in the interior of B^2.

Here the positive integer n is called the *(surface) braid index*. A surface braid is called *simple* if every branch point has branch index 2. A surface braid is *simple* if the covering in condition (1) above is simple. Throughout the paper we consider simple surface braids only.

Two surface braids are *equivalent* if they are isotopic under level-preserving isotopy relative to the boundary, where we regard the projection $p: B^2 \times D^2 \to D^2$ as a disk bundle over a disk.

2.2.2   *Closed surface braids*

Let $F \subset B^2 \times D^2$ be a surface braid with braid index n. Embed $B^2 \times D^2$ in 4-space standardly. Then ∂F is the unlink in the 3-sphere $\partial(B^2 \times D^2)$

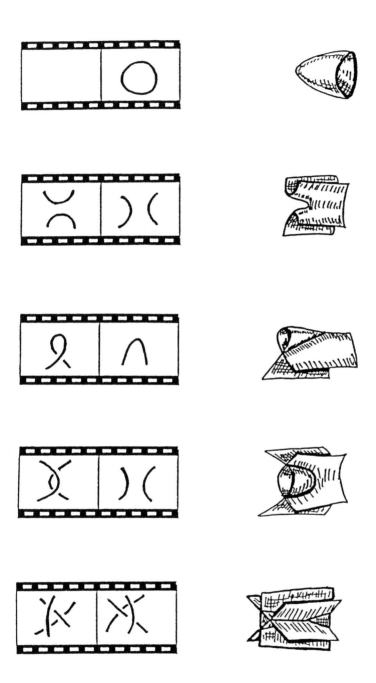

FIG. 2. The elementary string interactions

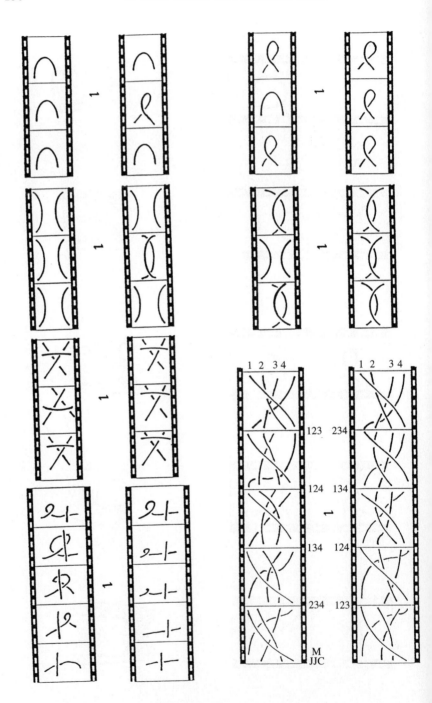

FIG. 3. The Roseman moves

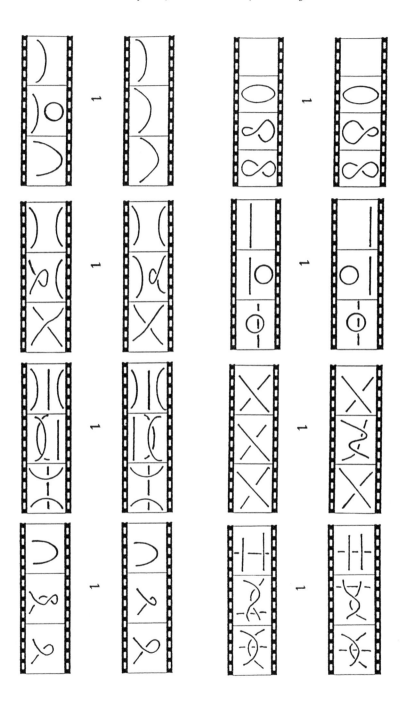

FIG. 4. The remaining movie moves

so that we can cap off these circles with the standard embedded 2-disks in 4-space to obtain a closed, orientable embedded surface in 4-space. The closed surface, $\hat{F}$ (called the *closure of F*), is defined up to ambient isotopy. A surface obtained this way is called a *closed surface braid*.

Viro and Kamada [23] have proven an Alexander theorem for surfaces. Thus, any orientable surface embedded in $\mathbf{R}^4$ is isotopic to a closed surface braid. Furthermore, Kamada [26] has recently proven a Markov theorem for closed surface braids in the PL category. However, in his proof he does not assume that the branch points are simple. An open problem, then, is to find a proof of the Markov theorem for closed surface braids in which each branch point is simple. Such a theorem would show a clear analogy between classical and higher-dimensional knot theory.

2.2.3 *Braid movies*

Write B^2 and D^2 as products of closed unit intervals: $B^2 = B_1^1 \times B_2^1$, $D^2 = D_1^1 \times D_2^1$. Factor the projection $p \colon B^2 \times D^2 \to D^2$ into two projections $p_1 \colon B_1^1 \times B_2^1 \times D^2 \to B_2^1 \times D^2$ and $p_2 \colon B_2^1 \times D^2 \to D^2$.

Let F be a surface braid. We may assume without loss of generality that p_1 restricted to F is generic. We further regard the D_2^1 factor in $B_2^1 \times D^2 = B_2^1 \times D_1^1 \times D_2^1$ as the time direction. In other words, we consider families $\{P_t = B_2^1 \times D_1^1 \times \{t\} : t \in D_2^1\}$. Except for a finite number of values of $t \in D_2^1$ the intersection $p_1(F) \cap P_t$ is a collection of properly immersed curves on the disk P_t.

Exceptions occur when P_t passes through branch points, or local maxima/minima of the double-point set, or triple points. We may assume that for any $t \in D_2^1$ there exists a positive ϵ such that at most only one of such points is contained in $\cup_{s \in (t-\epsilon, t+\epsilon)} P_s$. If one of such points is contained, the difference between $p_1(F) \cap P_s$, $s = t \pm \epsilon$, is one of the following: (1) smoothing/unsmoothing of a crossing (corresponding to a branch point), (2) type II Reidemeister move (corresponding to the maximum/minimum of the double-point set), or (3) type III Reidemeister move (corresponding to a triple point).

Except for such t, $p_1(F) \cap P_t$ is the projection of a classical braid. Furthermore, we can include crossing information by breaking the under-arc as in the classical case. Here the under-arc lies below the over-arc with respect to the projection p_1. To make this convention more precise we identify the image of p_1 with an appropriate subset in $B^2 \times D^2$. This idea was used by Kamada when he constructed a surface with a given chart. See [22] and also the next section.

By this convention we can describe surface braids by a sequence of classical braid words. Including another braid group relation ($\sigma_i \sigma_j = \sigma_j \sigma_i$, $|i - j| > 1$) we define *a braid movie* as follows.

A *braid movie* is a sequence $(b_1, \ldots, b_k)$ of words in $\sigma_i^{\pm 1}$, $i = 1, \ldots, n$, such that for any m, b_{m+1} is obtained from b_m by one of the following changes:

(1) insertion/deletion of σ_i^ϵ,
(2) insertion/deletion of a pair $\sigma_i^\epsilon \sigma_i^{-\epsilon}$,
(3) replacement of $\sigma_i^{\epsilon_1} \sigma_j^{\epsilon_2}$ by $\sigma_j^{\epsilon_2} \sigma_i^{\epsilon_1}$ where $|i - j| > 1$,
(4) replacement of $\sigma_i^\epsilon \sigma_j^\epsilon \sigma_i^\epsilon$ by $\sigma_j^\epsilon \sigma_i^\epsilon \sigma_j^\epsilon$ where $|i - j| = 1$.

These changes are called *elementary braid changes* (or EBCs, for short). This situation is illustrated in Fig. 5.

Two braid movies are *equivalent* if they represent equivalent surface braids.

2.3   Chart descriptions

A chart is an oriented labelled graph embedded in a disk, the edges are labelled with generators of the braid group, and the vertices are of one of three types:

(1) A *black vertex* has valence 1.
(2) A 4-valent vertex can also be thought of as a crossing point of the edges of the graph. The labels on the edges at such a crossing must be braid generators σ_i and σ_j where $|i - j| > 1$.
(3) *A white vertex* on the chart has valence 6. The edges around the vertex in cyclic order have labels $\sigma_i, \sigma_j, \sigma_i, \sigma_j, \sigma_i, \sigma_j$ where $|i - j| = 1$. Here, the orientations are such that the first three edges in this labelling are incoming and the last three edges are outgoing.

Consider the double point set of F under the projection $p_1|_F : B^2 \times D^2 \supset F \to B_2^1 \times D^2$ which is a subset of $B_2^1 \times D^2$. Then a chart depicts the image of this double point set under the projection $p_2 : B_2^1 \times D^2 \to D^2$ in the following sense:

(1) The birth/death of a braid generator occurs at a saddle point of the surface at which a double-point arc ends at a branch point. This projects to a black vertex on a chart.
(2) The projection of the braid exchange $\sigma_i \sigma_j = \sigma_j \sigma_i$ for $|i - j| > 1$ is a 4-valent vertex or a crossing in the graph.
(3) The 6-valent vertex corresponds to a Reidemeister type III move, and this in turn is a triple point of the projected surface.
(4) Maximal and minimal points on the graph correspond to the birth or death of $\sigma_i \sigma_i^{-1}$, and this is an optimum on the double-point set.

The relationships among surface projections, braid movies, and charts is depicted in Fig. 5.

2.3.1   *Reconstructing the surface braid from a chart*

A chart gives a complete description of a surface braid in the following way. Recall that the D_2^1 factor of D^2 is interpreted as a time direction. Thus

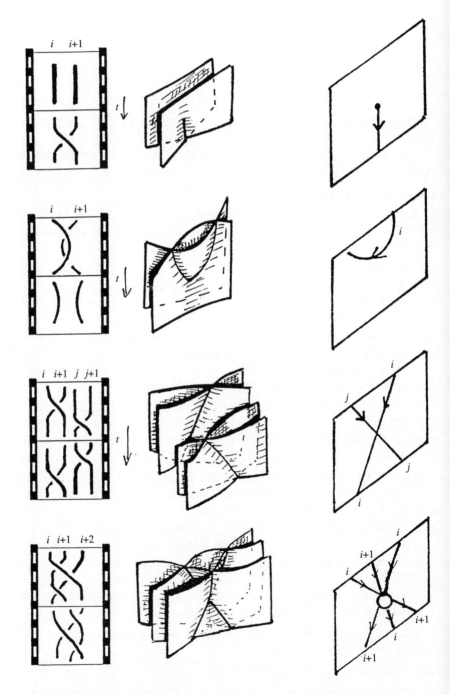

Fig. 5. Braid movies, charts, and diagrams

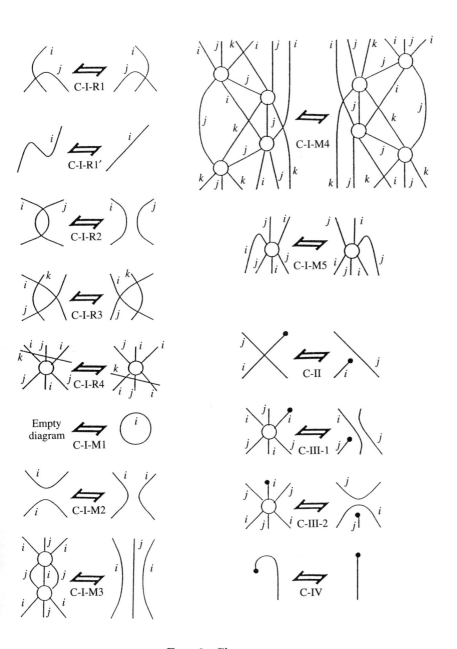

FIG. 6. Chart moves

the vertical axis of the disk points in the direction of increasing time. Each generic intersection of a horizontal line with the graph (generic intersections will miss the vertices and the critical points on the graph) gives a sequence of braid words. The letters in the word are the labels associated to the edges; the exponent on such a generator is positive if the edge is oriented coherently with the time direction, otherwise the exponent is negative. By taking a single slice on either side of a vertex or a critical point on the graph, we can reconstruct a braid movie from the sequence of braid words that results. Therefore, given a chart and a braid index, we can construct a surface braid and, indeed, a knotted surface by closing the surface braid.

Theorem 2.1   (Kamada) *Every surface braid gives rise to an oriented labelled chart, and every oriented labelled chart completely describes a surface braid.*

Furthermore, Kamada provided [25] a set of moves on charts. His list consists of three moves called CI, II, and III. From the braid movie point of view, we need to refine his list incorporating movie moves. Such generalized chart moves, yielding braid movie moves that will be discussed in the next section, are listed in Fig. 6.

2.4   Braid movie moves

The following 14 types of changes among braid movies are called *braid movie moves*:

(C-I-R1)

$$(\sigma_i, \sigma_i\sigma_j\sigma_j^{-1}, \sigma_j\sigma_i\sigma_j^{-1}) \leftrightarrow (\sigma_i, \sigma_j\sigma_j^{-1}\sigma_i, \sigma_j\sigma_i\sigma_j^{-1}),$$

where $|i - j| > 1$.

(C-I-R1′)

$$(\sigma_i, \sigma_i\sigma_i^{-1}\sigma_i, \sigma_i) \leftrightarrow (\sigma_i).$$

(C-I-R2)

$$(\sigma_i\sigma_j, \sigma_j\sigma_i, \sigma_i\sigma_j) \leftrightarrow (\sigma_i\sigma_j),$$

where $|i - j| > 1$.

(C-I-R3)

$$(\sigma_i\sigma_k\sigma_j, \sigma_k\sigma_i\sigma_j, \sigma_k\sigma_j\sigma_i, \sigma_j\sigma_k\sigma_i) \leftrightarrow (\sigma_i\sigma_k\sigma_j, \sigma_i\sigma_j\sigma_k, \sigma_j\sigma_i\sigma_k, \sigma_j\sigma_k\sigma_i),$$

where $|i - j| > 1, |j - k| > 1, |k - i| > 1$.

(C-I-R4)

$$(\sigma_k\sigma_i\sigma_j\sigma_i, \sigma_i\sigma_k\sigma_j\sigma_i, \sigma_i\sigma_j\sigma_k\sigma_i, \sigma_i\sigma_j\sigma_i\sigma_k, \sigma_j\sigma_i\sigma_j\sigma_k)$$

$$\leftrightarrow (\sigma_k\sigma_i\sigma_j\sigma_i, \sigma_k\sigma_j\sigma_i\sigma_j, \sigma_j\sigma_k\sigma_i\sigma_j, \sigma_j\sigma_i\sigma_k\sigma_j, \sigma_j\sigma_i\sigma_j\sigma_k),$$

where $|i - j| = 1$ and $|i - k| > 1, |j - k| > 1$.

(C-I-M1)

$$(\text{empty word}) \leftrightarrow (\text{empty word}, \sigma_i\sigma_i^{-1}, \text{empty word}).$$

(C-I-M2)

$$(\sigma_i\sigma_i^{-1}, \text{empty word}, \sigma_i\sigma_i^{-1}) \leftrightarrow (\sigma_i\sigma_i^{-1}).$$

(C-I-M3)

$$(\sigma_i\sigma_j\sigma_i, \sigma_j\sigma_i\sigma_j, \sigma_i\sigma_j\sigma_i) \leftrightarrow (\sigma_i\sigma_j\sigma_i),$$

where $|i - j| = 1$.

(C-I-M4)

$$(\sigma_i\sigma_j\sigma_k\sigma_i\sigma_j\sigma_i, \sigma_i\sigma_j\sigma_i\sigma_k\sigma_j\sigma_i, \sigma_j\sigma_i\sigma_j\sigma_k\sigma_j\sigma_i, \sigma_j\sigma_i\sigma_k\sigma_j\sigma_k\sigma_i,$$

$$\sigma_j\sigma_k\sigma_i\sigma_j\sigma_k\sigma_i, \sigma_j\sigma_k\sigma_i\sigma_j\sigma_i\sigma_k, \sigma_j\sigma_k\sigma_j\sigma_i\sigma_j\sigma_k, \sigma_k\sigma_j\sigma_k\sigma_i\sigma_j\sigma_k)$$

$$\leftrightarrow (\sigma_i\sigma_j\sigma_k\sigma_i\sigma_j\sigma_i, \sigma_i\sigma_j\sigma_k\sigma_j\sigma_i\sigma_j, \sigma_i\sigma_k\sigma_j\sigma_k\sigma_i\sigma_j, \sigma_k\sigma_i\sigma_j\sigma_k\sigma_i\sigma_j,$$

$$\sigma_k\sigma_i\sigma_j\sigma_i\sigma_k\sigma_j, \sigma_k\sigma_j\sigma_i\sigma_j\sigma_k\sigma_j, \sigma_k\sigma_j\sigma_i\sigma_k\sigma_j\sigma_k, \sigma_k\sigma_j\sigma_k\sigma_i\sigma_j\sigma_k),$$

where $k = j + 1 = i + 2$ or $k = j - 1 = i - 2$.

(C-I-M5)

$$(\sigma_j\sigma_i, \sigma_i^{-1}\sigma_i\sigma_j\sigma_i, \sigma_i^{-1}\sigma_j\sigma_i\sigma_j) \leftrightarrow (\sigma_j\sigma_i, \sigma_j\sigma_i\sigma_j\sigma_j^{-1}\sigma_j, \sigma_i^{-1}\sigma_j\sigma_i\sigma_j),$$

where $|i - j| = 1$.

(C-II)

$$(\sigma_j, \sigma_j\sigma_i, \sigma_i\sigma_j) \leftrightarrow (\sigma_j, \sigma_i\sigma_j),$$

where $|i - j| > 1$.

(C-III-1)

$$(\sigma_i\sigma_j, \sigma_i\sigma_j\sigma_i, \sigma_j\sigma_i\sigma_j) \leftrightarrow (\sigma_i\sigma_j, \sigma_j\sigma_i\sigma_j),$$

where $|i - j| = 1$.

(C-III-2)

$$(\sigma_j\sigma_j^{-1}, \sigma_j\sigma_i\sigma_j^{-1}, \sigma_i^{-1}\sigma_j\sigma_i) \leftrightarrow (\sigma_j\sigma_j^{-1}, \text{empty word}, \sigma_i^{-1}\sigma_i, \sigma_i^{-1}\sigma_j\sigma_i),$$

where $|i - j| = 1$.

(C-IV)

$$(\text{empty word}, \sigma_i^{-1}\sigma_i, \sigma_i) \leftrightarrow (\text{empty word}, \sigma_i).$$

Furthermore, we include the variants obtained from the above by the following rules:

(1) replace σ_i with σ_i^{-1} for all or some of i appearing in the sequences if possible,

(2) if $(b_1, \ldots, b_k) \leftrightarrow (b_1', \ldots, b_k')$ is one of the above, add $(b_k, \ldots, b_1) \leftrightarrow (b_k', \ldots, b_1')$,

(3) if $(b_1, \ldots, b_k) \leftrightarrow (b'_1, \ldots, b'_k)$ is one of the above, add $(c_1, \ldots, c_k) \leftrightarrow (c'_1, \ldots, c'_k)$ where c_j (resp. c'_j) is the word obtained by reversing the order of the word b_j (resp. b'_j).

(4) combinations of the above.

The braid movie moves are depicted in Figs 7 and 8. There is an overlap between the braid movie moves and the ordinary movie moves; that overlap is to be expected.

2.5 Definition (locality equivalence)

Let $(b_1, \ldots, b_n)$ and $(b'_1, \ldots, b'_n)$ be two braid movies that satisfy the following condition. There exists i, $1 \le i \le n$, such that

(1) $b_j = b'_j$ for all $j \ne i - 1, i, i + 1$,

(2) b_j (resp. b'_j), $j = i - 1, i, i + 1$, are written as $b_j = u_j v_j$ (resp. $b'_j = u'_j v'_j$) where u and v satisfy

   (2a) u_i (resp. v'_i) is obtained from u_{i-1} (resp. v'_{i-1}) by one of the elementary braid changes, say η, and $v_i = v_{i-1}$ (resp. $u'_i = u'_{i-1}$),

   (2b) v_{i+1} (resp. u'_{i+1}) is obtained from v_i (resp. u'_i) by one of the elementary braid changes, say ξ, and $u_i = u_{i+1}$ (resp. $v'_i = v'_{i+1}$).

Then we say that $(b'_1, \ldots, b'_n)$ is obtained from $(b_1, \ldots, b_n)$ by a *locality change* (or vice versa). If two braid movies are related by a sequence of locality changes then they are called *equivalent under locality*. Loosely speaking, equivalence under locality changes means that distant EBCs commute.

Theorem 2.2   [12] *Two braid movies are equivalent if and only if they are related by a sequence of braid movie moves and locality changes.*

It is interesting to note that in the braid movie point of view, the move that corresponds to a quadruple point gives a movie move corresponding to the permutohedral equation rather than to the tetrahedral equation. We will discuss these equations at some length in Section 3.

3   The permutohedral and tetrahedral equations

In [43], Zamolodchikov presented a system of equations and a solution to this system that corresponded to the interaction of straight strings in statistical mechanics. This system of equations is related to the Yang–Baxter equations, and this relationship suggests topological applications.

Baxter [1] showed that the Zamolodchikov solution worked. And in [2] higher-dimensional analogues were defined. In [37], these systems are interpreted as consistency conditions for the obstructions to the lower dimensional equations having solutions. Frenkel and Moore [16] produced a formulation and a solution of a variant of the Zamolodchikov tetrahedral equation. And Kapranov and Voevodsky [27] consider the solution of

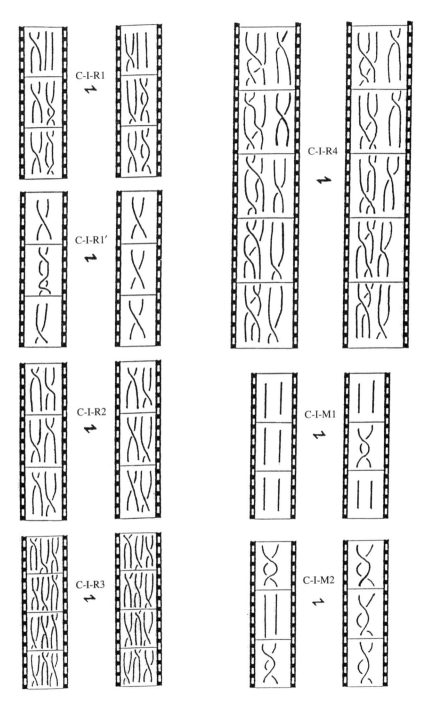

FIG. 7. The braid movie moves

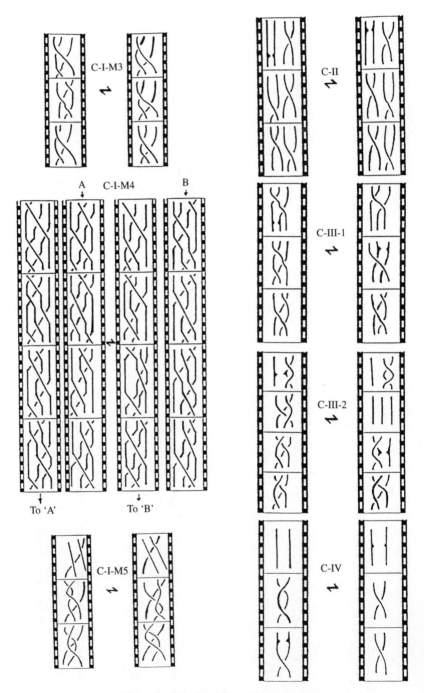

FIG. 8. The braid movie moves

the Zamolodchikov equation to be a key feature in the construction of a braided monoidal 2-category. In a parallel development, Ruth Lawrence [33] described the permutohedral equation and two others as natural equations for which solutions in a 3-algebra could be found.

There are other solutions of these higher-order equations known to us. In particular, we mention [13], [19], [28], and [31]. In [32], solutions are studied in relation to quasicrystals. We expect there are even more solutions in the literature and would appreciate being told of these.

Here we review these equations and our methods for solving them.

3.1 The tetrahedral equation

The Frenkel–Moore version of the tetrahedral equation is formulated as

$$S_{123}S_{124}S_{134}S_{234} = S_{234}S_{134}S_{124}S_{123} \qquad (3.1)$$

where $S \in \text{End}(V^{\otimes 3})$, V a module over a fixed ring k. Each side of the equation acts on $V^{\otimes 4} = V_1 \otimes V_2 \otimes V_3 \otimes V_4$, and S_{123}, for example, acts on V_4 as the identity. Notice here that eqn (3.1) has the nice property that each index appears three times on either side of the equation.

Compare this equation with the movie move that involves five frames on either side. On the left-hand movie, the elementary string interactions that occur are Reidemeister type III moves. Label the strings 1 to 4 from left to right in the first frame of the movie. (This index is one more than the number of strings that cross over the given string.) Then the Reidemeister type III moves occur in order among the strings (123), (124), (134), (234). Thus we think of the tensor S as corresponding to a type III Reidemeister move. The equation corresponds to the given movie move as these type III moves occur in opposite order on the right-hand side of the move. This correspondence was also pointed out by Towber.

3.1.1 *A variant*

There is a natural variant of this equation:

$$S_{124}S_{135}S_{236}S_{456} = S_{456}S_{236}S_{135}S_{124} \qquad (3.2)$$

where each side of the equation now acts on $V^{\otimes 6}$. In this equation each index appears twice in each side. This tetrahedral equation was also formulated in [27, 33, 37].

We review how to obtain eqn (3.2) from eqn (3.1) (cf. [27]). Define a set of pairs of integers among 1 to 4:

$$C(4,2) = \{12, 13, 14, 23, 24, 34\}$$

with the lexicographical ordering as indicated. There are six elements and the ordering defines a map $\xi \colon C(4,2) \to \{1, \ldots, 6\}$. The first factor of the left-hand side of the equation (3.1) is S_{123}. Consider the pairs of integers among this subscript: $\{12, 13, 23\}$ in the lexicographical ordering. Under

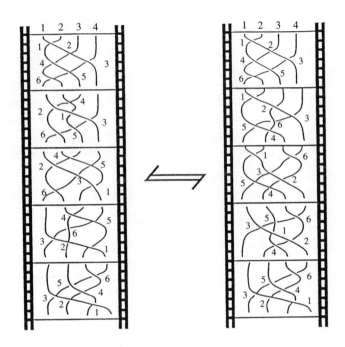

LHS

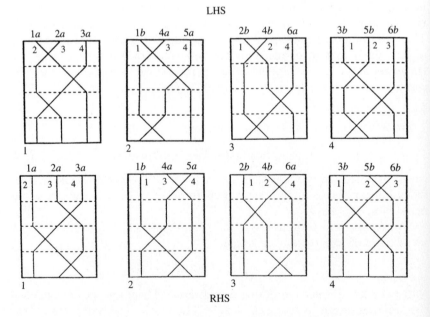

RHS

FIG. 9. Solving the Zamolodchikov equation

ξ these are mapped to $\{1, 2, 4\}$ which is the subscript of the first factor in eqn (3.2). The other factors are obtained similarly. For example, the second factor of (3.1) has the subscript 124 which yields pairs $\{12, 14, 24\}$ which are mapped under ξ to $\{1, 3, 5\}$ giving the second factor of (3.2). This is how we obtain (3.2) from (3.1).

Next we observe that eqn (3.2) is solvable: appropriate products of QYB solutions give solutions to (3.2). Let $V = W \otimes W$ where W is a module over which there is a QYB solution R: $R_{12} R_{13} R_{23} = R_{23} R_{13} R_{12}$. For given V give a and b as subscripts of W: $V_i = W_{ia} \otimes W_{ib}$ for $i = 1, \ldots, 6$. For $S = S_{123}$ acting on $V_1 \otimes V_2 \otimes V_3 = W_{1a} \otimes W_{1b} \otimes W_{2a} \otimes W_{2b} \otimes W_{3a} \otimes W_{3b}$ set

$$S_{123} = R_{1a2a} R_{1b3a} R_{2b3b}.$$

One can compute

$S_{124} S_{135} S_{236} S_{456}$

$$= R_{1a2a} R_{1b4a} R_{2b4b} R_{1a3a} R_{1b5a} R_{3b5b} R_{2a3a} R_{2b6a} R_{3b6b} R_{4a5a} R_{4b6a} R_{5b6b}$$
$$= R_{1a2a} R_{1a3a} R_{2a3a} R_{1b4a} R_{1b5a} R_{4a5a} R_{2b4b} R_{2b6a} R_{4b6a} R_{3b5b} R_{3b6b} R_{5b6b}$$
$$= R_{2a3a} R_{1a3a} R_{1a2a} R_{4a5a} R_{1b5a} R_{1b4a} R_{4b6a} R_{2b6a} R_{2b4b} R_{5b6b} R_{3b6b} R_{3b5b}$$
$$= R_{4a5a} R_{4b6a} R_{5b6b} R_{2a3a} R_{2b6a} R_{3b6b} R_{1a3a} R_{1b5a} R_{3b5b} R_{1a2a} R_{1b4a} R_{2b4b}$$
$$= S_{456} S_{236} S_{135} S_{124}.$$

Thus (3.2) is solvable.

This computation is a bit tedious algebraically. We obtained this result by considering the preimage of four planes of dimension 2 generically intersecting in 3-space. The movie version of such planes is depicted in Fig. 9. The numbering in the figure matches those described above. Specifically, the intersection points among planes 1 and 2 receive the number 1 since (12) is the first in the lexicographical ordering. Each plane is one of the strings as it evolves on one side of the movie of the corresponding movie move. The crossing points of the remaining planes give one side of the Reidemeister type III move on the preimage, and on the other side of the movie move the other side of the Reidemeister type III move appears. Thus the solution is a piece of bookkeeping based on this picture.

3.1.2 *Remarks*

In [8] we showed that solutions to the Frenkel–Moore equations could also produce solutions to higher-dimensional equations. All of the higher-dimensional solutions are obtained because the equations correspond to the intersection of hyperplanes in hyperspaces, and the self-intersection sets of these contain intersections of lower-dimensional strata.

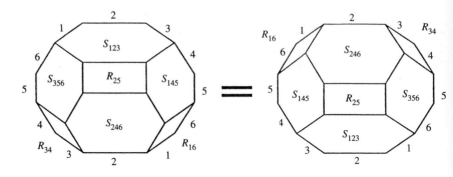

FIG. 10. The permutohedral equation

3.2 The permutohedral equation

This equation was studied in [27] and [33]:

$$S_{123}S_{145}R_{25}R_{16}S_{246}S_{356}R_{34} = R_{34}S_{356}S_{246}R_{25}R_{16}S_{145}S_{123}.$$

The subscripts indicate factors of vector spaces on which the tensor acts non-trivially, and it acts as the identity on the other factors. The tensors S and R live in $\text{End}(V^{\otimes 3})$ and $\text{End}(V^{\otimes 2})$, respectively, and both sides of the equation live in $\text{End}(V^{\otimes 6})$.

A pictorial representation is given in Fig. 10. This figure shows two sides of a permutohedron (truncated octohedron). The figure is the dual to the (C-I-M4) move on charts. Each hexagon is dual to a valence 6 vertex, and each parallelogram is dual to a crossing point on the graph. Thus this equation is a natural one to consider from the point of view of knotted surfaces, as it expresses one of the braid movie moves.

The equation expresses the equality among tensors that are associated to the faces on the '2 sides' of the permutohedron. Each hexagon represents the tensor S and each parallelogram represents R. Parallel edges receive the same number, and the numbers assigned to the boundary edges of hexagons and parallelograms determine the subscripts of the tensor.

Finally, observe that by setting R equal to the identity, the equation reduces to the variant of the tetrahedral equation given above.

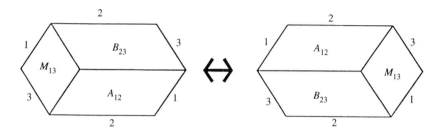

FIG. 11. Assigning tensors to the hexagons

3.2.1 *Solutions*

Suppose A, B, M, and $R \in \text{End}(V \otimes V)$ satisfy the following conditions:

$$AAA = AAA, \quad BBB = BBB \tag{3.3}$$
$$MMA = AMM, \quad BMM = MMB \tag{3.4}$$
$$MMA = AMM, \quad BMM = MMB \tag{3.5}$$
$$AMB = BMA \tag{3.6}$$
$$AMR = RMA, \quad RMB = BMR \tag{3.7}$$
$$ARM = MRA, \quad MRB = BRM \tag{3.8}$$
$$MAR = RAM \tag{3.9}$$
$$ARB = BRA, \quad BRA = ARB \tag{3.10}$$
$$BAR = RAB, \quad RAB = BAR \tag{3.11}$$
$$MRM = MRM \tag{3.12}$$

where the LHS (resp. RHS) of every relation has subscripts 12, 13, and 23 (resp. 23, 13, and 12) and lives in $\text{End}(V^{\otimes 3})$. The subscripts indicate the factor in the tensor product on which these operators are acting. For example, $AMM = MMA$ stands for $A_{12}M_{13}M_{23} = M_{23}M_{13}A_{12} \in \text{End}(V^{\otimes 3})$. We do not require the equalities with the subscripts of the RHS and LHS switched. In case those equalities are used, they are listed explicitly as in eqns (3.11) and (3.12).

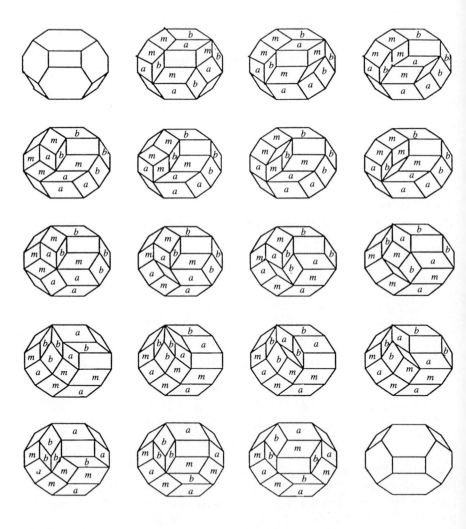

FIG. 12. The computational technique

Proposition 3.1   *If the conditions stated in eqns (3.3)-(3.12) hold, then*

$$S_{123} = A_{12}M_{13}B_{23}$$

is a solution to the permutohedral equation.

The proof is presented in detail in [11]. Here we include the illustrations that led to the proof. In this way we reinforce the diagrammatic nature of the subject.

In Fig. 11, we assign the product ABM to each of the S tensors. Then in Fig. 12, we indicate that if A, B, M, and R satisfy Yang–Baxter-like relations, then we can get from one side of the permutohedron to the other. To read this figure start at the upper left, travel left to right, then go directly down to the second row and read right to left. The zig-zag pattern continues.

Now in the solution given in (3.1) set $A = M = R$. Then the list of the conditions reduces to

$$RRR = RRR, \ BBB = BBB \qquad (3.13)$$
$$BRR = RRB, \ RRB = BRR \qquad (3.14)$$

with the same subscript convention as in eqns (3.3)-(3.12).

To get R-matrices that satisfy these conditions, we embedded quantum groups into bigger quantum groups. This idea was inspired by Kapranov and Voevodsky's idea of using projections of quantum groups to obtain solutions to Zamolodchikov's equation [27].

There exists a canonical embedding

$$\phi_i \colon U_{q_i}(sl(2)) \to U_q(sl(n))$$

corresponding to the ith vertex of the Dynkin diagram, where $q_i = q(\alpha_i, \alpha_i)/2$, $(-, -)$ is the canonical inner product, $\{\alpha_i\}_{i=1}^r$ is a basis of simple roots of the root system, and r is the rank (p. 173, [41]). Let R' (resp. B) be the universal R-matrix in $U_{q_i}(sl(2)) \otimes U_{q_i}(sl(2))$ (resp. $U_q(sl(n)) \otimes U_q(sl(n))$). Let $R = \phi_i(R')$ where i is fixed.

If we take a representation $f \colon U_q(sl(n)) \to \text{End}(V)$ for a vector space V over the complex numbers, both $f(R)$ and $f(B)$ satisfy the quantum Yang–Baxter equation ($RRR = RRR$, $BBB = BBB$) since $f(R) = f \circ \phi_i(R')$ is a representation of $U_{q_i}(sl(2))$ and R' is a universal R-matrix. Furthermore, in [11] we showed that the remaining relations held between R and B. In this way, we were able to construct solutions.

3.3   Planar versions

Simplex equations can be represented by planar diagrams when the subscripts consist of adjacent sequences of numbers. We can use planar diagrams to solve such equations.

3.3.1 *Planar tetrahedral equations*

One such equation is

$$S_{123}S_{234}S_{123}S_{234} = S_{234}S_{123}S_{234}S_{123}.$$

We regarded the corresponding planar diagram as the diagrams in the Temperley–Lieb algebra. Elements in the Temperley–Lieb algebra are represented by diagrams involving ∩ and ∪. Playing with such diagrams, we found a continuous family of solutions to the above equation expressed as a linear combination (with variables) in generators of the Temperley–Lieb algebra. Such generalizations of the Kauffman bracket seem useful in general.

3.3.2 *The planar permutohedral equation*

It was observed by Lawrence [33] that the permutohedral equation becomes the following equation:

$$r_{34}s_{123}s_{345}r_{56}r_{23}s_{345}s_{123}$$

$$= s_{456}s_{234}r_{45}r_{12}s_{234}s_{456}r_{34} \tag{3.15}$$

after multiplying by permutation maps $s = P_{13}S$, $r = PR$.

We found some new and interesting solutions to this version by using the Burau representation of the braid group, and these were presented in [11]. Some of the solutions were non-invertible, and another class of solutions reflected the non-injectivity of the Burau representation.

4 2-categories and the movie moves

At the Isle of Thorns Conference in 1987, Turaev described the category of tangles and laid the foundations for his paper [42]. The objects in the category are in one-to-one correspondence with the non-negative integers. And the morphisms are isotopy classes of tangles joining n dots along a horizontal to m dots along a parallel horizontal. Such a tangle, then, is a morphism from n to m. A week later in France similar ideas were being discussed by Joyal [21]. All of the known generalizations of the Jones polynomial can be understood as representations on this category. Consequently, there is hope that the yet to be found higher-dimensional generalizations can be found as representations of the 2-category of tangles that we will describe here.

4.1 Overview of 2-categories

In a category, there are objects. For any two objects in a category there is a set of morphisms. A 2-category has objects, morphisms, and morphisms between morphisms (called 2-*morphisms*). The 2-morphisms can be composed in essentially two different ways: either they can be glued along common 1-morphisms, or they can be glued along common objects.

Obviously, they can be glued only if they have either a morphism or an object in common.

Just as the fundamental problem in an ordinary category is to determine if two morphisms are the same, the fundamental problem in a 2-category is to determine if two 2-morphisms are the same. The geometric depiction of a 2-category consists of dots for objects, arrows between dots for morphisms, and polygons (where 'poly' means two or more) for 2-morphisms. Thus, by composing 2-morphisms we obtain faces of polyhedra, and an equality among 2-morphisms is a solid bounded by these faces.

4.2 The 2-category of braid movies

This section is an interpretation of Fischer's work in the braid movie scheme. Here we give a description of the 2-category associated to n-string 2-braids. There is one object, the integer n, and this is identified with n points arranged along a line. The set of morphisms is the set of n-string braid diagrams (without an equivalence relation of isotopy imposed). Two diagrams related by a level-preserving isotopy of a disk which keeps the crossings are identified, but for example two straight lines and a braid diagram represented by $\sigma_i \sigma_i^{-1}$ are distinguished. Equivalently, a morphism is a word in the letters $\sigma_1^{\pm 1}, \ldots, \sigma_{n-1}^{\pm 1}$. A generating set of 2-morphisms is the collection of EBCs.

More generally, consider the 2-category in which the collection of objects consists of the natural numbers $\{0, 1, 2, \ldots\}$; each number n is identified with n dots arranged along a horizontal line. The set of morphisms from n to m when $m \neq n$ is empty, and the set of morphisms from n to n is the set of n-string braid diagrams (or equivalently words on the alphabet $\{\sigma_1^{\pm 1}, \ldots, \sigma_n^{\pm 1}\}$). A tensor product is defined by addition of integers, and 0 acts as an identity for this product. A 2-morphism is a surface braid that runs between two n-string braid diagrams. Equivalently, a 2-morphism can be represented by a movie in which stills differ by at most an EBC.

We define a braiding which is a 1-morphism from $n + m$ to $m + n$, by braiding n strings in front of m strings; that is, the braid diagram represented by the braid word

$$(\sigma_n \sigma_{n+1} \cdots \sigma_{n+m-1}) \cdot \cdot (\sigma_{n-1} \sigma_n \cdots \sigma_{n+m-2}) \cdot \cdots \cdot (\sigma_1 \sigma_2 \cdots \sigma_m).$$

Assertion 1   *With the tensor product, identity, and braiding defined above, we can define all the data (various 2-morphisms) in terms of braid movies such that the 2-category of braid movies defines a braided monoidal 2-category.*

Sketch of Proof In Figs 13 and 14 a set of movie moves to ribbon diagrams are depicted. These are illustrations of the Kapranov–Voevodsky axioms as interpreted in our setting.

In the middle of figures polytopes are depicted. These are two exam-

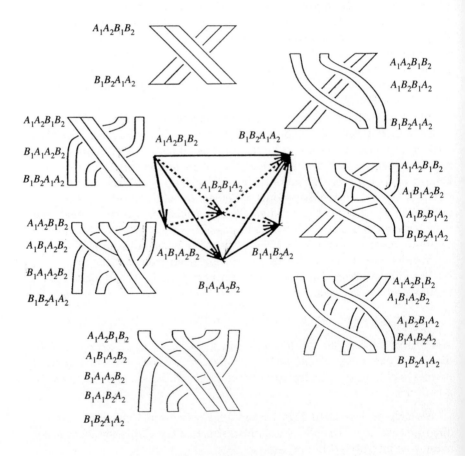

FIG. 13. A ribbon movie move

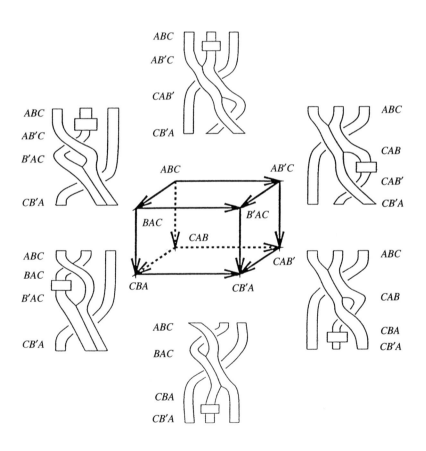

FIG. 14. Another ribbon movie move

ples of axioms in [27] that are to be satisfied by 2-morphisms. Letters represent objects where tensor product notation is abbreviated. Edges in these polytopes are 1-morphisms. They are in turn represented by braid words. The ribbon diagrams surrounding the polytopes represent braid movies. Ribbons represent parallel arcs as in the classical case. Then each 1-morphism, a braid word, is represented by ribbon diagrams to simplify movie diagrams. Therefore each ribbon diagram represents a composition of edges in the polytopes. A 2-morphism is a face in the polytopes, and changes between two ribbon diagrams. The polytopes are relations satisfied by compositions of 2-morphisms. In ribbon diagrams, this means that two ways of deforming ribbon diagrams have to be equal. There are two ways to deform (by braid movie moves) one (the top ribbon diagram) to the other (the bottom): by going along the left hand side of the polytope, or the right hand side. The conditions say that these two ways are equal.

Each frame in a movie can be written as a product of braid generators while each 2-morphism can be decomposed as a sequence of EBCs. These EBCs can be written in terms of charts. Then two ribbon movies are equivalent if and only if the resulting charts are chart equivalent. Thus we can check the equality by means of chart moves.

Alternatively, we can decompose the polytopes into simpler pieces that correspond to the polytopes of the braid movie moves.   □

4.2.1 *Remark*

The existence of a braided monoidal 2-category is not enough to conclude that one has an invariant of 2-knots. In addition, to the braiding $R: A \otimes B \rightarrow B \otimes A$, we need a braiding $\bar{R}: B \otimes A \rightarrow A \otimes B$ that also gives a 2-morphism from $\bar{R} \circ R$ to the identity 1-morphism on $A \otimes B$. If R is thought of as crossing the strings of A over those of B, then $\bar{R}$ crosses B's strings over A's. Since moves C-I-M1 and C-I-M2 hold, these 2-morphisms are 2-isomorphisms.

4.2.2 *Fischer's result*

Fischer's dissertation gives an axiomatization of the 2-category that is associated to regular isotopy classes of knotted surface diagrams. That is, he defines this 2-category, and he shows that it is the free, rigid, braided, semistrict monoidal 2-category (FRBSM2-cat) generated by a set. Fischer's theorem means that if one has an RBSM2-cat then one automatically has constructed an invariant of regular isotopy classes of knotted surface diagrams.

5 Algebraic interpretations of braid movie moves

Our motivation in defining braid movies and their moves was to get a more algebraic description of knotted surfaces. In particular, it is hoped that this description can be defined and studied in a more general setting:

for example, for any group. Also, it is desirable if there is an algebraic-topological description of braid movies generalizing the fact that a classical braid represents an element of the fundamental group of the configuration space. In this section, we work towards this goal; we observe close relations between braid movie moves and various concepts in algebra and algebraic topology.

5.1 Identities among relations and Peiffer elements

In this section we recall definitions of various concepts for group presentations following [3], where details appear.

5.1.1 *Identities among relations*

Let G be a group. A *presentation* of G is a triple $\langle X; R, w \rangle$ such that R is a set, $w: R \to F$ is a map, where F is the free group on X, and G is isomorphic to $F/N(w(R))$, where $N(\ )$ is the normal closure.

Let H be the free group on $Y = R \times F$; let $\theta: H \to F$ denote the homomorphism $\theta(r, u) = u^{-1}w(r)u$. Then $\theta(H) = N(w(R))$. An *identity among relations* is any element in the kernel of θ.

5.1.2 *Peiffer transformations*

A Y-sequence is a sequence of elements $\mathbf{y} = (a_1, \ldots, a_n)$ where $a_i = (r_i, u_i)^{\pm 1}$ is a generator (or its inverse) of H for $i = 1, 2, \ldots, n$. The following operations are called *Peiffer transformations*:

(1) A *Peiffer exchange* replaces the successive terms $(r_i, u_i), (r_{i+1}, u_{i+1})$ with either

$$(r_{i+1}, u_{i+1}), \ (r_i, u_i u_{i+1} w(r_{i+1}) u_{i+1})$$

or

$$(r_{i+1}, u_{i+1} u_i^{-1} w(r_i)^{-1} u_i), \ (r_i, u_i).$$

(2) A *Peiffer deletion* deletes an adjacent pair (a, a^{-1}) in a Y-sequence.

(3) A *Peiffer insertion* inserts an adjacent pair (a, a^{-1}) into a Y-sequence.

(4) A *Peiffer equivalence* is a finite sequence of exchanges, deletions, and insertions to a Y-sequence performed in some order.

5.1.3 *Pictures for presentations*

Let $\langle X, R, w \rangle$ denote a group presentation. A *picture over the presentation* consists of

(1) A disk D with boundary ∂D.

(2) A collection $\Delta_1, \ldots, \Delta_k$ of disjoint disks in the interior of D that are labelled by elements of R.

(3) A collection of disjoint oriented arcs labelled by elements of X that either form simple closed loops or join points on the boundaries of two (or possibly one) of the Δ.

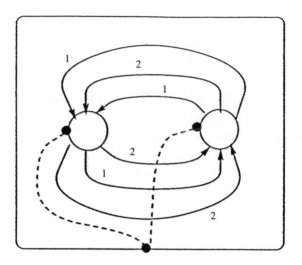

FIG. 15. Pictures for group presentations

(4) A base point for each of the Δ and one for the big disk D. The base point on a Δ is distinct from the endpoints of the connecting arcs.

(5) Finally, starting from the base point and reading around the boundary of any Δ in a counterclockwise fashion, the labels on the edges that are encountered should spell out either the relation R or its inverse that is associated to the small disk. An incoming edge is interpreted as having a positive exponent, and an outgoing edge has a negative exponent.

In [3] *a spherical picture* is defined as a picture in which none of the edges touch the outer boundary. Figure 15 illustrates a spherical picture. The dotted arcs indicate how to obtain a Y-sequence from a picture. The Y-sequence obtained from a spherical picture is *an identity sequence* in the sense that the product $\theta(a_1) \cdots \theta(a_n) = 1$ in the free group.

A picture determines a map of a disk into BG, the classifying space of the group with given presentation, that is defined up to homotopy, and by transversality if such a map is given then there is a picture: the little Δ disks are the inverse images of regular neighborhoods of points at the center of the disks that correspond to relators. The arcs are the inverse images of interior points on the edges of the complex.

From a picture a Y-sequence can be obtained. Namely, for each Δ disk a 'dotted' arc is chosen that connects the base point of the small disk to that of the larger disk. These arcs do not intersect in their interiors. Furthermore, the dotted arcs are chosen to be transverse to the arcs that interconnect the disks. To each dotted arc we associate a pair $(r^{\pm}, u)$, where $r \in R$ and $u \in F$. The element u is the word in the free group corresponding to the intersection sequence of the dotted arc with connecting arcs. The exponent on a letter is determined by an orientation convention. The element r is the intersection sequence on the boundary of Δ read in a counterclockwise fashion. The collection of dotted arcs is ordered in a counterclockwise fashion at the base point, and this gives the ordering of the Y-sequence.

In [3], it is noted that an elementary Peiffer exchange corresponds to rechoosing a pair of successive arcs from the base points of a pair of Δ to the base point of D. Thus pictures can be used to study Y-sequences and their Peiffer equivalences. They provide a diagrammatic tool for the study of groups from an algebraic-topological viewpoint.

Furthermore, they [3] define deformations of pictures. There are three types of moves to pictures. The first type is to insert or delete small circles labelled by the generating set. This is a direct analogue of the C-I-M1 move on charts. Secondly, surgery between arcs with compatible labels and orientations is analogous to the C-I-M2 move. Third is an insertion or deletion of a pair of small disks (Δ) labelled with the same relator, but oppositely oriented, joined together by arcs corresponding to generators that appear in the relator. This is an isolated configuration disjoint from other components of the graph in the picture, but corresponds to C-I-M3 and C-I-R2 (combined with C-I-M1, C-I-M2).

5.2 Moves on surface braids and identities among relations

In this subsection we observe which moves on surface braids correspond to Peiffer transformations defined above, and what the others correspond to.

5.2.1 *Moves corresponding to Peiffer transformations*

Braid movie moves C-I-M3 and C-I-R2 are used to perform Peiffer insertions and deletions as noted in the previous section. Peiffer exchanges correspond to locality equivalences because ordering the vertices is tantamount to choosing a height function.

5.2.2 *Moves corresponding to 2-cycles*

The chart moves C-I-R3, C-I-R4, and C-I-M4 are identities among relations that correspond to non-degenerate 2-cycles in the Cayley complex for the standard presentation of the braid group. The correspondence is given by taking duals. The cube, the hexagonal prism, and the permutohedron, respectively, are the dual 2-cycles in the Cayley complex.

5.2.3 *Other C-I moves*

These correspond to rechoosing the dotted arcs in the homotopy classes relative to the endpoints.

For example, C-I-R1$'$ looks like straightening up a cubic curve. Thus if the x-axis in the figure is a part of a 'dotted arc', then this move corresponds to rechoosing a dotted arc so that it intersects the edge with two less intersection points. Other moves can be interpreted similarly.

This rechoosing in turn corresponds to changing the words u_i in the terms (r_i, u_i) of the Y-sequences, since dotted arcs correspond to these words.

In summary, we interpret charts without black vertices as null homotopies of the second homotopy group of the classifying space. Let G be a braid group B_n on n strings and BG its classifying space constructed from the standard presentation. Specifically, BG is constructed as follows: start with one vertex, add n oriented edges corresponding to generators, attach 2-cells corresponding to relators along the 1-skeleton, add 3-cells to kill the second homotopy group of the 2-skeleton thus constructed, and continue killing higher homotopies to build BG. Note that pictures are defined for maps of disks to BG as in the case of maps to the Cayley complexes by transversality.

Proposition 5.1   *Let C be a chart of a surface braid S without black vertices. Note here that S is trivial (i.e. S is an unlinked collection of n 2-spheres). Then there is a map f from a disk to BG, $f \colon (D^2, \partial D^2) \to (BG, vertex)$, such that the picture corresponding to f is the same (isotopic as a planar graph) as C. Since $\pi_2(BG) = 0$, this picture is deformed to the empty picture by a sequence of picture moves. This is the same as a sequence of CI moves realizing an isotopy of surface braids from S to the trivial surface braid (whose chart is the empty diagram). In other words, CI moves of isotopy of surface braids correspond to picture moves of $\pi_2(BG)$ $(= 0)$.*

Furthermore, this deformation is interpreted as identities among relations. In particular, braid movie moves correspond to Peiffer transformations, or 3-cells of BG, or rechoosing the 'dotted arcs' in pictures.

5.3   Symmetric equivalence and moves on surface braids

In this section we consider moves on group elements that are conjugates of generators. This generalizes Kamada's definition of symmetric equivalence for charts. These moves correspond to C-II and C-III moves on charts which are the moves involving black vertices.

5.3.1   *Symmetric equivalence and charts*

Fix a base point on the boundary of the disk on which the chart of a surface braid lives. Suppose there are black vertices. Consider a small disk B centered at a black vertex v. Pick a point d on the boundary of B which misses the edge coming out of v. Connect d to the base point by a simple arc which intersects the chart transversely. (More precisely, the arc misses the vertices and crossings of the chart and intersects the edges transversely.)

Starting from the base point, travel along the arc and read the labels as it intersects the edges. Then go around v along ∂B counterclockwise, and go back to the base point. This yields a word w in braid generators of the form of a conjugate of a generator: $w = Y^{-1}\sigma_i^\epsilon Y$, where σ_i is the label of the edge coming from v, $\epsilon = \pm 1$ depends on the orientation of this edge, and Y is the word we read along the arc.

Then before/after the C-II move, the word thus defined changes as follows:

$$Y^{-1}\sigma_i^\epsilon Y \leftrightarrow Y^{-1}\sigma_j^{-\delta}\sigma_i^\epsilon\sigma_j^\delta Y,$$

where δ also denotes ± 1 and $|i - j| > 1$. Similarly a C-III move causes the word change

$$Y^{-1}\sigma_i^\epsilon Y \leftrightarrow Y^{-1}\sigma_j^{-\delta}\sigma_i^{-\delta}\sigma_j^\delta\sigma_i^\delta\sigma_i^\delta Y,$$

where $|i - j| = 1$, $\epsilon = \pm 1$, and $\delta = \pm 1$.

Kamada defined *symmetric equivalence* between words that are conjugates of generators in the braid group. The equivalence relation reflects the changes above. He proved that two such words are symmetrically equivalent iff they represent the same element in the braid group, and then he used this theorem to prove that C moves suffice. (Kamada's definition of symmetric equivalence is different from the above. We indicated the word changes that correspond to C-II and C-III moves.)

5.3.2   Chart moves that involve black vertices

Charts that have black vertices can be interpreted as pictures in the sense of [3] as follows. Consider a chart with black vertices. Cut the disk open along the arcs defined above to obtain a picture in which the boundary of the disk is mapped to the product of conjugates of generators. This gives a picture associated to a chart that has black vertices.

The chart moves C-II and C-III have the following interpretation as moves on pictures. The words on the boundary of the disk change as indicated above.

The picture represents a map from the disk into BG where G is the braid group. Either of these moves corresponds to sliding the disk by homotopy across the polygon representing the given relation.

5.3.3   Generalizations of Wirtinger presentations

Generalizing Kamada's theorem, we consider Wirtinger presentations of groups. Let $G = \langle X : R \rangle$ be a finite Wirtinger presentation. Thus each of the relators is of the form $r = y^{-1}V^{-1}xV$, where $x, y \in X$ are generators and V is a word in generators. Let A be the set of all the words in X (i.e., elements in the free group on X) of the form $w = Y^{-1}xY$, where $x \in X$ and Y is a word in X. Define *symmetric equivalence* on A as follows:

(1) Two words $Y^{-1}xY$ and $Z^{-1}xZ$ are symmetrically equivalent if Y and Z represent the same element in G.

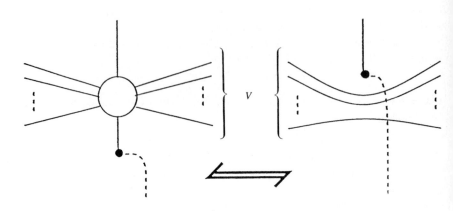

FIG. 16. Generalized chart move

(2) Two words $Y^{-1}yY$ and $Y^{-1}V^{-1}xVY$ are symmetrically equivalent if $r = y^{-1}V^{-1}xV \in R$ where $x, y \in X$.

Two words in A are symmetrically equivalent iff they are related by a finite sequence of the above changes.

In terms of pictures, if we allow univalent vertices in pictures, and if we define arc systems connecting univalent vertices to a base point on the boundary of the disk, then we can diagrammatically represent the above two changes in terms of pictures. They are depicted in Fig. 16. Such a generalization of pictures will be discussed next.

We generalize pictures for group presentations to include univalent vertices. To make homotopy-theoretic sense out of such pictures, consider small disks centered at each univalent vertex and simple, mutually disjoint arcs connecting them to a base point. More precisely, our data consist of:

(1) a picture on a disk where univalent vertices are allowed,

(2) a small disk centered at each univalent vertex with a specified point on the boundary of each disk which misses the edge coming from the univalent vertex,

(3) a system of mutually disjoint simple arcs such that for each univalent vertex a simple arc from this system connects the specified point on the boundary of the small disk to a fixed base point on the boundary of the (big) disk where the picture lives. These arcs miss vertices of the picture and intersect the edges transversely.

Let $(a_1, \ldots, a_n)$ be the system of arcs defined above. Starting from the base point, read the labels of the edges intersecting a_1 when we trace it towards the univalent vertex. Go around the vertex along the boundary of the small disk and come back to the base point to get a word w_1. Then we trace a_2 up to a_n to get a sequence $(w_1, \ldots, w_n)$ of words. Each word is in a conjugate form $w_i = Y_i^{-1} x_i Y_i$ $(i = 1, \ldots, n)$, where x_i is the label of the edge coming from the ith univalent vertex. Now we require that the product $w_1 \cdots w_n$ is 1 in G, the group with which we started. Call this a *generalized picture*.

This has the following meaning as explained in the case of braid groups above. Given a generalized picture, remove the interior of small disks around each univalent vertex. Then define a smooth map from the boundary big disk cut open along 'dotted' arcs connecting small disks to the base point, to the 1-skeleton of BG. This edge path represents the product $w_1 \cdots w_n$ in $G = \pi_1(BG)$ which was assumed to be the identity. Therefore the map on the boundary of the cut-open disk extends to a map of the big disk. The transverse preimages of points in 1-cells (resp. 2-cells) gives edges (resp. vertices) of the rest of the picture.

Here we remark that Kamada's theorem on symmetric equivalence seems to be generalized to other groups. Kamada used Garside's results [17], and Garside gave other groups to which his arguments can be applied.

5.4 Concluding remarks

We have indicated relations among braid movie moves, charts and classifying spaces of braid groups, and null homotopy disks in such spaces. We further discussed potential generalizations to Wirtinger presented groups. We conclude this section with remarks on these aspects.

In the preceding section we discussed null homotopy disks in classifying spaces in the presence of (generalized) black vertices. However, we do not want to allow cancelling black vertices since if we did charts represent just null homotopies in the case of braid groups and we instead want something highly non-trivial. Thus we need restrictions on homotopy classes of maps from cut-open disks to classifying spaces to be able to define nontrivial homotopy classes to represent surface braids. Keeping base point information is such an example. More natural interpretations are expected in terms of other standard topological invariants.

By fixing a height function on the disk where pictures are defined, we can define sequences of words for Wirtinger presented groups generalizing braid movies. Furthermore we can generalize braid movie moves to such sequences for Wirtinger presented groups by examining moves on pictures in the presence of height functions. The moves in this case would consist of (1) the moves that correspond to Peiffer equivalences and the moves on pictures described in [3], (2) the moves that are dependent on the choice of height functions, (3) the moves that correspond to 2-cycles in the Cayley complex associated to the presentation, (4) the moves that push black vertices through Wirtinger relations. The algebraic significance of such sequences of words and relations among them deserves further study, since this is a natural context in which the braid movie moves appear.

Acknowledgements

We thank the following people for valuable conversations and correspondence: John Baez, Dan Silver, Dan Flath, John Fischer, Seiichi Kamada, Ruth Lawrence, Dennis McLaughlin, and Kunio Murasugi. Some of the material here was presented at a special session hosted by Yong-Wu Rong and at a conference hosted by David Yetter and Louis Crane. Financial support came from Yetter and Crane and from John Luecke. Cameron Gordon's support (financial and moral) is especially appreciated.

Bibliography

1. Baxter, R. J. (1983) On Zamolodchikov's Solution of the Tetrahedron Equations, *Commun. Math. Phys.* **88**, 185–205; Reprinted in Jimbo, M., (1989) *Yang–Baxter Equation in Integrable Systems,* World Scientific Publishing, Singapore.

2. Bazhanov, V. V. and Stroganov, Yu. G. (1989) Chiral Potts Models as a Descendant of the Six-Vertex Model, in Jimbo, M., (1989) *Yang–Baxter Equation in Integrable Systems,* World Scientific Publishing, Singapore.

3. Brown, R. and Huebschmann, J. (1982) Identities among Relations, in Brown, R. and Thickstun, T. L. (1982) *Low-dimensional Topology,* London Math. Soc. Lecture Note Series **48**, pp. 153–202.

4. Carter, J. Scott and Saito, Masahico, (1991) Syzygies among Elementary String Interactions in Dimension 2+1, *Lett. Math. Phys.* **23**, 287-300.

5. Carter, J. Scott and Saito, Masahico (1992) Canceling Branch Points on Projections of Surfaces in 4-Space, *Proc. AMS* **116**, 229–237.

6. Carter, J. Scott and Saito, Masahico (1992) Planar Generalizations of the Yang–Baxter equation and Their Skeins, *J. Knot Theory Ramifications* **1**, 207–217.

7. Carter, J. Scott and Saito, Masahico (1993) Reidemeister Moves for Surface Isotopies and Their Interpretation as Moves to Movies, *J. Knot Theory Ramifications* **2**, 251–284.

8. Carter, J. Scott and Saito, Masahico (1993) On Formulations and Solutions of Simplex Equations, Preprint.

9. Carter, J. Scott and Saito, Masahico (1993) Diagrammatic Invariants of Knotted Curves and Surfaces, Preprint.

10. Carter, J. Scott and Saito, Masahico (1993) A Diagrammatic Theory of Knotted Surfaces, to appear in Baddhio, R. and Kauffman, L. (1993) *The Conference Proceedings for AMS Meeting #876, on Knots and Quantum Field Theory,* World Scientific Publishing, Singapore.

11. Carter, J. Scott and Saito, Masahico (1993) Some New Solutions to the Permutohedron Equation, Preprint.

12. Carter, J. Scott and Saito, Masahico (1993) Braids and Movies, Preprint.

13. Ewen, H. and Ogievetsky, O. (1992) Jordanian Solutions of Simplex Equations, Preprint.

14. Fischer, John (1993) Braided Tensor 2-categories and Invariants of 2-knots, Preprint.

15. Fox, R. H. (1962) A Quick Trip Through Knot Theory, in Cantrell, J. and Edwards, C. H. (1962) *Topology of Manifolds,* Prentice Hall.

16. Frenkel, Igor and Moore, Gregory (1991) Simplex Equations and Their Solutions, *Commun. Math. Phys.* **138**, 259–271.

17. Garside, F. A. (1969) The Braid Group and Other Groups, *Q. J. Math. Oxford* **20**, 235–254.

18. Golubitsky, Martin and Guillemin, Victor (1973) *Stable Mappings and Their Singularities,* GTM **14**, Springer Verlag.

19. Hietarinta, J. (1992) Some Constant Solutions to Zamolodchikov's Tetrahedron Equations, Preprint.

20. Jones, V. F. R. (1985) A Polynomial Invariant for Knots and Links via von Neumann Algebras, *Bull. AMS* **12**, 103–111. Reprinted in Kohno, T. (1989) *New Developments in the Theory of Knots,* World Scientific Publishing, Singapore.

21. Joyal, A. and Street, R. (1988) Braided Monoidal Categories, Mathematics Reports 86008, Macquarie University.

22. Kamada, Seiichi (1992) Surfaces in $\mathbf{R}^4$ of Braid Index Three are Ribbon, *J. Knot Theory Ramifications* **1**, 137–160.

23. Kamada, Seiichi (1993) A Characterization of Groups of Closed Orientable Surfaces in 4-space, to appear in *Topology.*

24. Kamada, Seiichi (1993) 2-Dimensional Braids and Chart Descriptions, Preprint.

25. Kamada, Seiichi (1993) 2-jigen braid no chart hyoji no douchi-henkei ni tsuite (On Equivalence Moves of Chart Descriptions of 2-Dimensional Braids), Preprint (in Japanese).

26. Kamada, Seiichi (1993) Generalized Alexander's and Markov's Theorems in Dimension Four, Preprint.

27. Kapranov, M. and Voevodsky, V. (1989) Braided Monoidal 2-

Categories, 2-Vector Spaces and Zamolodchikov's Tetrahedra Equations (first draft). Preprint.

28.   Kapranov, M. and Voevodsky, V. (1992) 2-Categories and Zamolodchikov Tetrahedra Equations, Preprint.

29.   Kawauchi, A., Shibuya, T., and Suzuki, S. (1982) Descriptions of Surfaces in Four-space, *Math. Semin. Notes Kobe Univ.* **10**, 75–125.

30.   Kawauchi, A., Shibuya, T., and Suzuki, S. (1983) Descriptions of Surfaces in Four-space, *Math. Semin. Notes Kobe Univ.* **11**, 31–69.

31.   Kazhdan, D. and Soibelman, Y. (1993) Representations of Quantized Function Algebras, 2-Categories and Zamolodchikov Tetrahedra Equation, Preprint.

32.   Korepin, V. E. (1987) Completely Integrable Models in Quasicrystals, *Commun. Math. Phys.* **110**, 157–171.

33.   Lawrence, Ruth (1992) On Algebras and Triangle Relations, in Mickelsson, J. and Pekonen, O. (1992) *Topological and Geometric Methods in Field Theory,* World Scientific Publishing, Singapore, pp. 429–447.

34.   Libgober, A. (1989) Invariants of Plane Algebraic Curves via Representations of the Braid Groups, *Invent. Math.*, 9525–9530.

35.   Lin, X.-S. (1990) Alexander–Artin–Markov Theory for 2-links in $\mathbf{R}^4$, Preprint.

36.   Lomonaco, Sam (1981) The Homotopy Groups of Knots I. How to Compute the Algebraic 2-Type, *Pacific J. Math.* **95** 349–390.

37.   Maillet, J. M. and Nijhoff, F. W. (1989) Multidimensional Integrable Lattices, Quantum Groups, and the D-Simplex Equations, Inst. for Non-Linear Studies preprint 131.

38.   Roseman, Dennis (1981) Reidemeister-Type Moves for Surfaces in Four Dimensional Space, Preprint.

39.   Rudolph, Lee (1983) Braided Surfaces and Seifert Ribbons for Closed Braids, *Commun. Math. Helv.* **58**, 1–37.

40.   Rudolph, Lee (1983) Algebraic Functions and Closed Braids, *Topology* **22**, 191–202.

41.   Soibelman, Y. (1991) The Algebra of Functions on a Compact Quantum Group, and its Representations, *Leningrad Math. J.*, **2**.

42.   Turaev, V. (1988) The Yang–Baxter Equation and Invariants of Links, *Invent. Math.* **92**, 527–553. Reprinted in Kohno (1989) *New Developments in the Theory of Knots,* World Scientific Publishing, Singapore.

43.   Zamolodchikov, A. B. (1981) Tetrahedron Equations and the Relativistic S-Matrix of Straight-Strings in $2+1$-Dimensions, *Commun. Math. Phys.* **79**, 489–505, Reprinted in Jimbo, M. (1989) *Yang-*

Baxter Equation in Integrable Systems, World Scientific Publishing, Singapore.